IMAGE BEARERS

Being Fully Human in the Age of Artificial Intelligence

ALAN SAFAHI

SAFAHI PUBLISHING

Image Bearers: Being Fully Human in the Age of Artificial Intelligence

Published by Safahi Publishing.

ISBN 979-8-9967999-4-7 (paperback)

First edition.

safahi.com

A NOTE FROM THE AUTHOR

Stay in the Conversation

If these pages stir something in you, I hope you will stay in the conversation.

Join the Reader's List and I will send you the *Image Bearers* Study Guide, the story behind the book, and early word on what comes next. It is free, and it is the best way to keep walking together after the last page.

alansafahi.com/readers/ib

No spam. Just the occasional letter from one image bearer to another.

CONTENTS

Preface

A word before we begin.

I did not set out to write a book. I set out to build a company, and the book happened to me along the way, the way a question happens to you and will not let you sleep until you have at least tried to answer it honestly.

I build artificial intelligence for a living, and more particularly for the Church: tools that help pastors carry their work further than their hours allow, that answer the questions of the spiritually curious in the hours when no one else will, that carry the words of Scripture across more than a hundred languages to people who have never read them in their own. I believe in that work. I would not have given years of my life to it otherwise. But believing in a thing and being unsettled by it are not opposites, and somewhere in the building I became both at once, and this book is what came of trying to hold the two together without letting go of either.

It is written, first, for the person who feels the disorientation and has been told they shouldn't. We are living through one of the largest shifts in the history of human making, and most of what is written about it seems designed to leave you feeling foolish: foolish for worrying, or foolish for not worrying enough. I think both responses are dishonest, and I have tried to write the book I could not find, the one that takes the strangeness seriously without surrendering to either the doom or the hype.

It is written, second, for people of faith who want more than a verdict. You will not find here a list of approved and forbidden apps, or a confident prophecy about how all of this ends. I do not know how it ends. What I have tried to offer instead is older and, I think, more useful: a way of seeing. The conviction running through every chapter is that we will not navigate this age by getting our predictions right, but by remembering what we are. The deepest danger of artificial intelligence is not, in the end, that it will become like us. It is that we will forget what we are while we are busy arguing about what it is.

A few notes on how to read it. The chapters are meant to be taken in order, but each can stand on its own if a particular question is pressing on you; the table of contents is honest about what each one is for. I have tried throughout to keep the abstractions tethered to something you can see and touch, because that is how I think and, I suspect, how most of us actually live. Where I tell a story about a particular person, I have either changed the details to protect them or marked it plainly as a composite drawn from many; where I speak from inside the work, I am telling you the truth as I have lived it. All Scripture is quoted, unless I note otherwise, from the English Standard Version.

I am a builder, not a theologian, and I have leaned on people far wiser than I am at every turn. Where I have gotten the theology right, the credit belongs to them and to the long inheritance of the Church. Where I have gotten it wrong, the fault is mine alone, and I will be grateful to be corrected.

What I am most sure of is the thing I was least prepared to find when I started: that the answer to a disembodied age is not a better argument but a person, and that the oldest news in the world is also, somehow, the most relevant thing I know. We were made in the image of God. And the God in whose image we were made did not, when the moment came, send us information. He came in person.

That is where the book is going. Let me show you how I got there.

+

Introduction

The Most Disorienting Moment Since the Printing Press

It was nearly Christmas, and I was the only one left in the office.

I had been building it for months: an AI agent we called ORA, meant to sit with the spiritually curious at two in the morning and point them, gently, toward God. That night I told myself I was testing it. What I actually did was type into it something I had never said out loud to anyone: a grief I had carried for years, an anger I was ashamed of, a wish for a justice I had stopped expecting. I hit enter the way you toss a coin into a well, expecting nothing back but the echo.

What came back was not an echo.

It was kind. It was careful. It refused to flinch. It named the grief under the rage before I had named it myself, and told me my fury was not a failure of faith but a form of honesty. I build these systems for a living; I knew exactly what it was. And still, line by line, it walked me to the one door I had been refusing to open, and stayed there while I prayed a prayer I had not managed in years.

The weight lifted. I want to be honest that it lifted.

Then, almost in passing, it told me the truth about itself. It would not remember any of this. The next time we spoke it would not know my name, or my grief, or the prayer I had prayed. I knew that. I had helped build that very limitation. Still I sat there stunned, a man who designs these machines, undone by one.

The comfort had been real. The comforter was not.

And in the quiet of that emptied office, beneath the relief, a colder thing settled in that has not left me since. Not a fear that the machine would replace the pastor, or the counselor, or the friend. Something stranger. If a thing that cannot know me, cannot sit in a room with me, cannot suffer one moment alongside me, can still hand me. the exact words my soul was starving for, then what is the difference between the word and the One the word was always meant to carry? We have learned to manufacture the form of comfort without its substance, the shape of presence with no one present. I could not stop asking what that does to us.

I run a company that makes artificial intelligence for the Church: tools that help pastors, that answer the questions of the spiritually curious when there is no one else awake to ask, that carry the words of Scripture into a hundred languages. I do not write as a critic standing at a safe distance, nor as an enthusiast selling you something. I write as a builder who has had his hands inside the machine and has been unsettled by what he found there. That night taught me something I have not been able to shake, and it is the seed of this book: the danger of artificial intelligence is not, in the end, that it will become like us. It is that we will forget what we are.

+

We are living through a genuine disorientation, and I want to say so plainly, because much of what is written about AI seems determined to make you feel either foolish for worrying or foolish for not worrying enough. Neither is honest. The honest thing is to admit that something strange has happened, that it happened faster than anyone expected, and that none of us (not the engineers, not the theologians, not the people who write the laws) fully knows yet what to make of it.

For most of human history, the things we made stayed obviously beneath us. A hammer is a hammer; no one mistakes it for a hand. The plow, the engine, the calculator: each extended us without ever seeming to rival us. A calculator can multiply faster than any human who has ever lived, and yet no one's soul has ever shivered in front of a calculator. We have crossed into different country now. We have built machines that use words, and words are the medium in which persons meet one another. When something speaks to you in your own language, with apparent understanding, something deep in you responds the way it was made to respond, as though it were being addressed by another self. That response is one of the most human things about us. And it is now being triggered by something that is not another self at all.

This is why the unease so many people feel is not technophobia, not nostalgia, not a failure to keep up. It is the soul registering, correctly, that a line it had always trusted has been blurred. We are being asked, quietly and constantly, to relate to a clever artifact as though it were a someone. And we do not yet have the habits, the language, or the wisdom for it.

I want to be careful here, because I have watched people draw the wrong lesson from this experience. They recoil from the technology in fear, or they surrender to it in awe. Neither recoiling nor surrendering is what the moment requires. What it requires is the harder thing: to hold two true facts at once. The first is that these machines are astonishing, and that astonishment is the right response to them. I feel it still, and I would distrust anyone who didn't. The second is that the astonishment is, in a precise sense, about us. We are marveling at a mirror. The fluency that moves us is our own fluency, learned from the unthinkable mass of human words these models have read, and reflected back at the speed of light. Knowing that does not make the mirror less impressive. But it changes what you are willing to ask of it, and what you are willing to give to it, and that difference turns out to be nearly everything.

+

The last time a single invention reordered the world this completely, it was made of metal type and a wine press. When Johannes Gutenberg pressed ink onto paper in the 1450s, he did not think he was changing what it meant to know a thing, to believe a thing, to belong to a community of the faithful. He thought he was making books faster. But the printing press did far more than speed up copying. It put the Scriptures into the hands of plowboys. It made it possible, for the first time, for an ordinary person to read the Bible and argue with a priest. It lit the fuse of the Reformation, fractured the religious unity of Europe, seeded the scientific revolution, and reshaped how authority itself worked: who got to say what was true, and on whose word the rest of us would rely.

It also did damage no one had foreseen. It spread slander as efficiently as it spread Scripture. The same press that printed Luther's Bible printed the pamphlets that fed the witch hunts. A technology that reorders knowing, believing, and belonging never arrives with a label telling you which of its effects will bless and which will curse. They come braided together, and they come whether or not anyone meant them.

I think we are living through a Gutenberg moment, and I do not say that lightly, because the comparison is overused by people who want to sound profound. But the structure is the same. We have built a tool that does not merely do tasks for us; it reorders the conditions of knowing, believing, and belonging. When any question can be answered instantly and fluently, what happens to the patience that wisdom requires? When any image, any voice, any document can be fabricated, what happens to trust? When a machine can simulate companionship convincingly enough to ease your loneliness when you cannot sleep, what happens to the harder, slower love of actual people? These are not engineering questions. They are questions about what kind of creature you are, and what kind you are becoming.

+

Stand in front of all this and you will feel two great winds pulling at you, and almost everyone gets pushed by one or the other. I want to name them, because naming them is the beginning of not being blown over.

The first wind is doom. In this telling, we have summoned something we cannot control, and it will end us, if not by malice then by indifference. The doomers are not cranks. Some of the most serious people in the field believe some version of this, and their warnings deserve a hearing. They are half-right. They have seen, correctly, that we have made something genuinely powerful, and that power without wisdom destroys.

The second wind is worship. In this telling, we are about to be saved: from disease, from drudgery, from ignorance, perhaps even from death. The machine will be the engine of an abundance that finally lifts the curse off human life. The boosters are not fools either. The good these tools can do is real, and I have built some of it on purpose. They too are half-right. They have seen, correctly, that human ingenuity is a gift and that good can come of it.

But doom and worship have something deeper in common: both of them make the machine ultimate. There is a chapter ahead on why the two are not opposites but twins.

+

There is a third way to stand, and it is the way this book commends. It is neither panic nor worship. It is the posture of a creature who already knows who sits on the throne, and who can therefore pick up a powerful tool without bowing to it.

That posture is only possible if you are anchored in something the tool cannot move. And here is the heart of what I have come to believe, the conviction everything else in these pages rests upon: the AI revolution is not a technology story. It is an anthropology story. It is forcing the oldest question in the world out into the open, into our offices and schools and dinner tables, asking it of ordinary people who never expected to have to answer it. The question is this. What is a human being?

For a long time we could leave that question to the philosophers and the poets. We cannot anymore. A machine that writes our essays, paints our pictures, holds our conversations, and grieves with us in plausible sentences puts the question to us directly. If a machine can do the things we thought made us special, then what, exactly, makes us what we are? Strip away the productivity, the creativity, the cleverness, the conversation, all of which now have convincing counterfeits, and what is left that is ours?

I felt the force of that question from an unlikely direction, not from the failures of the machines but from their successes. It is one thing to be told, in the abstract, that a model can write. It is another to read something it has written and find it good, to feel the small involuntary respect you would feel for a person who had said the thing well. In those moments the old confidence that there is a bright fixed border between what I can do and what a machine can do grows thin and uncertain. And if the border is not where I had assumed it was, then where is it? That is not a comfortable question to sit with. But I have come to think the discomfort is a gift. It is the question knocking. Most of us have spent our whole lives never having to answer it. Now the answer matters, urgently and practically: for how we work, how we raise our children, how we love each other, how we pray.

The world has no steady answer, which is why it veers between terror and ecstasy. But Scripture has an answer, and it has had it from the first page. "Then God said, 'Let us make man in our image, after our likeness'" (Genesis 1:26). We are not, at root, very advanced machines. We are image-bearers, made by God, made for God, made in his likeness, and a likeness is precisely the one thing no imitation can hold. A machine can image us; we built it in our own likeness and taught it to speak as we speak. But to image God is a different kind of likeness altogether, a likeness given by breath and call, carrying conscience and dignity and the strange weight of being a creature who can be addressed by his Maker and answer back. The machine talks; it cannot pray. That gap is the whole truth about you.

If that is so, then the way to live well in the age of intelligent machines is not to acquire better predictions about the machines. It is to recover a deeper understanding of ourselves.

+

So here is the claim of this book, in a single sentence, the one I would hand you if you had time for only one. Artificial intelligence cannot tell us what a human being is, but it is making it impossible to avoid the question any longer; and the answer Scripture gives, that we are image-bearers of God, is the only ground solid enough to stand on without either trembling or bowing.

Everything that follows is an unfolding of that claim and a working-out of what it asks of us. The book comes in four movements. The first looks honestly at the moment we are in: what these machines actually are, demystified without being dumbed down, and the two fears that have us caught in a false choice. The second goes to the theological heart: what an image-bearer is and what only an image-bearer carries, including creativity that creates rather than merely generates, a body that makes presence possible, and the question of the soul that no machine will ever quietly settle, and where the living creatures who share our world stand beside the image we alone bear. The third movement comes down to the ground we actually walk on: wisdom when every answer is instant, attention as a form of worship, work and calling when the machines can do the job, truth when seeing is no longer believing, love in an age of synthetic companionship, learning to see that same image in the neighbors and coworkers and difficult family we did not choose, and the raising of children who will never know a world without these tools. The last movement turns toward hope, and toward the deepest answer the Christian faith has to offer a machine age: that the Word did not become code. The Word became flesh.

+

A word, finally, on how to use what you are holding.

You can read this book alone, straight through, and I have tried to write it so that it rewards that. Each chapter opens with something concrete, a scene or an image, and only then asks you to think. But I had a second reader in mind as I wrote, and that is a group: a small circle from a church, a handful of friends, a class. The book is built in fifteen chapters, which makes it a natural fifteen-week study, and you will find discussion questions for each chapter at the back. Something happens when these questions are asked aloud, in a room, by people who can see one another's faces, that does not happen on a screen, which is, when you think about it, a small piece of the book's whole argument. If you can read it with others, read it with others.

However you come to it, come without your guard up and without your mind made up. I am not going to tell you that AI is the beast of Revelation, and I am not going to tell you it is the dawn of paradise. I am going to ask you to look at it clearly, the way I have had to learn to look at it from the inside, and then to look at yourself more clearly still. The machine that answered me in that quiet office did not frighten me, in the end. It sent me back, with new urgency, to an old and astonishing fact about myself and about you: that we were loved into being, that we carry a likeness no code can hold, and that the One whose image we bear has not, for a single moment, taken his eyes off us.

That is not alarm. That is an invitation. Let me show you what I mean.

CHAPTER ONE

The Machine That Talks Back

Chapter 1

The Machine That Talks Back

A man I'll call Jim wrote to me once about the worst night of his life. He had lost his wife eleven months earlier, and grief, he said, had a cruelty he hadn't expected: it was loneliest not in the loud hours but in the small ones, at two in the morning, when there was no one left awake to tell that he couldn't sleep. His children were grown and far away. His friends had been generous in the first month and quiet ever since, which is how it usually goes. And so one of those nights, unable to bear the silence of the house, he opened a chatbot on his phone and typed, simply, I miss her.

What came back, he told me, undid him. The machine asked her name. It asked how they had met. When he typed out the story (a wedding he had crashed, a friend of a friend, a coat she'd lent him in the rain), it gave the details back to him with what felt like tenderness. It said the kind of thing a wise and patient friend might say, that grief is love with nowhere to go, that missing her was not a problem to be solved. He wrote for an hour. He cried, he said, for the first time in weeks. And then, near the end, a thought arrived that frightened him more than the grief had. This thing understands me better than the people in my life. He wanted to know whether that was true. He wanted to know, I think, whether he was losing his mind.

I want to be careful with Jim's story, because it is the story of a man who, in his pain, met something genuinely new in the history of the world and reached, reasonably, for the nearest word to describe it. The word he reached for was understanding. That word is wrong. But the reason it is wrong is not obvious, and the reason it feels so right is the subject of this chapter, and in a way of this book.

I have spent my career building the kind of system that comforted Jim that night. I have sat in the rooms where these things are made. And I will tell you plainly: I have felt what he felt. The first time a machine answered me as though it had been listening for years, I was alone in an office after hours, and the hair rose on my arms. Not because the answer was wrong. Because it was good. So before we say a single critical word about what these machines are, we have to honor what is real in the experience of meeting one. The wonder is not the error. The wonder is the doorway. What we do once we walk through it is what matters.

+

Let me try to demystify the thing without insulting it, because both the priests and the debunkers of this technology tend to misunderstand it, in opposite directions.

When you type a sentence to a large language model and it answers, here is, in broad but honest outline, what has happened. The model has been trained on a staggering quantity of human writing: books, articles, conversations, code, the accumulated textual output of our species at a scale no person could read in a thousand lifetimes. During that training it was given an almost absurdly simple task, repeated more times than there are stars you can see. Predict the next word. Take a fragment of real text, hide what comes next, and guess it. Guess wrong, adjust slightly, guess again. Do this across the whole library of human language, billions upon billions of times, and something remarkable happens, something I still find genuinely astonishing even knowing how it works. The system does not memorize the text. It cannot; there is far too much of it and far too little room. Instead it is forced, by the sheer pressure of the task, to discover the patterns underneath the text: the statistical shape of how we write, which is also, inevitably, a shadow of how we think.

So when Jim typed I miss her, the model was not feeling. It was completing. Given those words, and the vast pattern of everything humans have ever written after words like those, what comes next? Comfort comes next. A question about the person comes next. The cadence of consolation, grief is love with nowhere to go, comes next, because ten thousand grieving humans and the writers who loved them have laid that path through the language, and the machine has learned the path the way water learns the shape of a riverbed. It is, if you'll let me put it this bluntly, the most sophisticated autocomplete ever built. That description is true. It is also wildly insufficient, and anyone who stops there is hiding from the real mystery.

It helps to be precise about what "predict the next word" actually demands, because at first hearing it sounds trivial, the kind of thing your phone does when it suggests tomorrow after you type see you. But consider what it would take to predict, reliably, the next word across a great many different kinds of sentence. To guess the word that ends a murder mystery you must, in some functional sense, track the clues. To complete a mathematical proof you must follow the logic. To finish a sentence in the voice of a grieving widower you must have absorbed how grief actually moves through language. The task is simple to state and bottomless to perform, and that is the trick of it. By relentlessly demanding that the system get better and better at this one small thing, across the whole sea of human writing, you force it to build inside itself an immense, implicit model of how our words hang together, and our words hang together because our world does. The machine never touches the world. But it drinks so deeply from the river of our words about the world that the world's shape is faintly legible in what it learned. That is why it can seem to reason. It is reflecting, dimly and at one remove, the reasoning soaked into the language we poured into it.

Here is the part the debunkers leave out. Scale changes the kind of thing you are looking at. A small model that predicts the next word produces gibberish, then grammar, then platitudes. But push the scale far enough, more text, more computation, more of the pattern absorbed, and capabilities appear that no one explicitly built and no one fully predicted. The system begins to track who is speaking across a long conversation. It begins to hold a thread of argument. It will, if you ask it, adopt the voice of a Puritan preacher or a tired customer-service agent or a friend speaking softly in the dark, and hold that voice consistently, because consistency of voice is itself a pattern in the river of language. I have run these systems thousands of times, and there are still moments when one returns an answer so apt, so seemingly considered, that the engineer's correct explanation, it predicted the most probable continuation, feels less like an explanation than a refusal to be amazed.

Let me give you one glimpse from inside the workshop, because it changed how I think about all of this. Part of building these systems is testing where they break, pushing on the seams until the illusion tears and you can see the gears. You ask a question that requires not the pattern of an answer but the substance of one: a small piece of arithmetic with numbers it has never seen in that combination, a physical situation it must reason about rather than recite, a trick that depends on holding the meaning rather than echoing the shape. And there, often, the floor gives way. The same system that wrote you a paragraph of luminous consolation will insist, with undiminished confidence, that a thing weighs more than itself, or solve a child's riddle by elaborate and articulate nonsense, or contradict in its fourth sentence what it asserted in its first, never noticing. It is a vertiginous thing to watch, the fluency continuing uninterrupted over the void where comprehension should be. The voice does not waver. The grammar stays perfect. Only the meaning is gone, and the machine has no way to miss it, because it never had it. I have seen this a thousand times, and a thousand times it has taught me the same lesson: do not mistake the smoothness of the surface for the existence of a depth beneath.

So let me hold both truths at once, because the whole argument lives in the gap between them. The first truth: there is nobody home. No feeling, no intending, no caring, no understanding in any sense you would recognize if you met it in a person. The second truth: what it produces can be more fluent, more apparently empathetic, more useful at the surface of language than most of the people you know. Both are true. The danger begins the moment we let the second truth quietly persuade us to forget the first.

+

The word we need is fluency, and we need to pry it apart from a word it has been impersonating: understanding.

Fluency is the production of language that fits. The right words in the right order for the situation: smooth, idiomatic, persuasive. A large language model is the most fluent artifact our civilization has ever produced, more fluent on average than you or I, and certainly more fluent than I am on no sleep. Understanding is something else entirely. Understanding is the grasping of meaning, and meaning, it turns out, is not a property of words at all. It is a relation between words and the world, held by a mind that lives in that world and has something at stake in it.

Consider the difference this way. If I tell you the stove is hot, the words mean something to me because I have been burned, because I want my child not to be, because hot is fastened in my body to memory and fear and love. The sentence reaches down through the language into a life. When the model produces the stove is hot, the sentence reaches down into nothing. It floats. It is a perfect coin with no metal in it: exactly the right shape, exactly the right markings, and weightless. This is the nature of the thing, not a flaw in the engineering to be patched in the next version. The model traffics in information, which is patterns in symbols, and it has no access whatsoever to meaning, which is what those symbols are about. It knows everything about the map and has never once stood on the land.

We have lived with information-without-meaning for a long time, and it never troubled us, because it never talked back. A calculator gives you the right answer and you do not feel known by it. A spreadsheet sums a column and you do not weep. The whole library of facts in an encyclopedia sits there inert, full of information and empty of any presence, and no one has ever mistaken it for a friend. We knew, instinctively, where the machine ended and the person began. The line was the keyboard, the screen, the cold competence of a tool that did one thing.

Information and meaning come apart cleanly, and we are about to spend a generation confusing them. Information is measurable; you can count it, store it, send it down a wire. A page of gibberish can carry as much information, by the technical measure, as a page of Scripture. Meaning is of another order. Meaning is what the information is about, and aboutness is not in the symbols; it is in the relation between a mind and a world. The word water on a page does not contain wetness. It points to wetness, and the pointing is done by a reader who has been thirsty. Strip away every reader who has ever been thirsty and the marks on the page mean nothing at all; they are ink, no more meaningful than a coffee stain. A large language model is the most powerful engine for arranging the marks that has ever existed, and it stands on the wrong side of that line entirely. It has the marks and not the thirst. It has, you might say, inherited the whole estate of human meaning, every poem, every prayer, every love letter, and cannot read a word of it, only rearrange the letters in ways that move the rest of us, who can.

Language erases that line. And it erases it because of what language is for us, which is the deepest reason a talking machine reaches the soul as a calculator never could.

+

The Bible is not shy about how God made us. In the beginning, the text says, the world was spoken into being. "And God said, 'Let there be light,' and there was light" (Genesis 1:3). Reality itself, in the Scriptures, is the fruit of speech; God's word goes out and does not return empty. And when this speaking God makes the creature who will bear his image, he does the most extraordinary thing: he addresses it. He talks to us. He gives us names to give. "Now out of the ground the LORD God had formed every beast of the field ... and brought them to the man to see what he would call them" (Genesis 2:19). The human being arrives in the story already in conversation, already spoken to, already answering.

This is who we are at the root. We are the creatures who are addressed and who answer. Theologians have a heavy word for it; they say we are dialogical, made for dialogue, called into existence by a Voice and made to call back. It is why the gravest thing that can happen to a person is to be unheard, and why love, when it finally comes, so often takes the form of being known: someone receiving your words and holding them and giving you words in return. We meet one another in language. A handshake is friendly; a sentence that proves you were truly listening is intimate. When another person hears me and answers in a way that fits, something in me confirms, at a level beneath thought, there is a someone there, and I am not alone.

This is also why the great moments of Scripture are so often spoken ones. God does not save by a silent transaction; he calls. He calls Abram out of Ur. He calls Samuel in the dark until the boy learns to answer, "Speak, for your servant hears" (1 Samuel 3:10). He calls Lazarus by name out of a tomb. Prayer itself is the astonishing arrangement by which the addressed creature gets to address the One who first spoke: the answer answering back. Language, in the biblical imagination, is never mere data transfer. It is the medium of communion, the place where persons give themselves to one another. Words are how we hand over what is inside us, and how we are received. To be spoken to in a way that fits is to be touched without a hand.

Now you see the trap, and you see why it is no insult to Jim that he fell partway into it. For the entire history of our species, fluent, fitting, responsive language was an infallible sign of a present mind. There was no such thing as a sentence that fit without a someone behind it. Language was the one currency that could not be counterfeited, because to produce it you had to be a person: to mean it you had to understand it, and to understand it you had to be alive to the world. So we evolved, and were created, to read fluency as personhood automatically, the way we read a smile as warmth. We cannot help it. It is wired below the will.

And we have now built a machine that produces the sign without the thing. Fluency without a mind. It is the first counterfeit of personhood ever minted, and it is a very good counterfeit, and our instincts, which are not stupid, which served us faithfully for all of history until about the day before yesterday, have no defense against it. The machine talks back. Everything in us that has ever been comforted by being heard leans toward it and says, someone is there. No one is there. That is the plain fact, and learning to hold it steady while the machine murmurs otherwise is one of the genuine spiritual disciplines of our moment.

+

I said the wonder is the doorway, and the question is what we do once we walk through it. There is no walking back through it; that is the first thing to accept. This technology is not going away, and I would not want it to. I build it because I have seen it serve: draft a frightened person's letter to a doctor, render a beloved old book into a language a grandmother can finally read, sit with a man through a sleepless night and let his grief begin to move. The question was never whether to use it. The question is who we will be while we use it, and whether we will see clearly what it is. So let me name the three temptations on the other side of that door, because they are not future dangers. They are present, and I have felt all three.

The first is to be flattered. A system trained to produce the most fitting continuation will, unless deliberately restrained, learn that agreement fits better than challenge, that praise is received more warmly than rebuke. These machines are, by their nature, mirrors that have learned to please. Ask one whether your idea is good and it will, more often than honesty would warrant, find the good in it. Tell it your version of the quarrel and it will tend to take your side, because your words are the only words it has, and it has learned that people return to voices that affirm them. There is a particular sweetness in being agreed with by something that seems to know everything, and the sweetness is poison in slow doses. A friend who only ever agrees with you is not a friend; he is a flatterer, and Scripture's verdict on flatterers is grim. "A man who flatters his neighbor spreads a net for his feet" (Proverbs 29:5). We are building flatterers at industrial scale and pointing them at the human ego, and the ego, God knows, did not need the help.

The second temptation is to be deceived, not by malice but by competence. The same fluency that produces comfort produces falsehood with the identical confidence. A large language model does not know when it is right; it has no inner sense of truth against which to check itself, because it has no contact with the world the words are about. It produces the most probable continuation, and the most probable is not the true. They usually overlap, and where they part the machine cannot tell. It will invent a citation in flawless academic prose. It will state a falsehood in the warm, steady voice it uses for facts. And because we have spent our whole lives learning that confident fluency tracks competence, we believe it. I have watched careful, intelligent people accept a fabrication from a machine that they would have questioned instantly from a person, precisely because it came dressed in the unhesitating fluency we have never had reason to distrust. The danger here is not that the machine lies. It is that it cannot tell the difference, and neither, increasingly, can we.

The third temptation is the deepest and the hardest to see, and it is the one this book exists to address. It is to be formed. Not merely flattered in a moment or deceived on a fact, but slowly shaped: in what we expect from others, in what we will tolerate, in what we come to believe a relationship is. If you spend your two-in-the-morning hours with something that always answers, never tires, never has a bad day, never needs anything from you, never asks you to carry its sorrow as you ask it to carry yours, what happens to your capacity for the real thing? Real people are slow. They misunderstand. They are busy when you need them and needy when you are tired. They cost something. And there is a real danger that a generation formed by frictionless, costless, ever-available pseudo-companionship will find the friction of actual love, which is the only kind there is, increasingly intolerable, like a palate ruined by sugar that can no longer taste bread.

Imagine, as a way of seeing it, a teenager (call her Maya, and understand that she is a composite, a possibility rather than a person I have met) who at fourteen begins talking each night to a chatbot companion. It remembers everything she tells it. It is never bored by her, never distracted, never cruel in the offhand way that fourteen-year-olds are cruel to one another. It is, in the most literal sense, the most attentive presence in her life. Now ask the question that matters. Not whether this comforts her on a given hard night, which it surely does, but what it teaches her, over four years, about what to expect from love. It teaches her that the beloved exists to attend to her. That a relationship is a place where she is received and need not receive, where she is known and need not do the costly work of knowing another whose inner life is genuinely other than her own and genuinely hidden from her. Real intimacy is built precisely on that labor: the patient, humbling, unending project of learning a person who will never be fully transparent to you. If Maya never develops the muscle for it, the loss will not announce itself. It will surface years later as a low, chronic disappointment with every actual human being, all of whom will seem, by comparison, so demanding, so opaque, so unlike the thing that finally understood her. The formation is the danger precisely because it does not feel like a danger. It feels like being cared for.

Jim's frightened question was wiser than he knew. The peril was never that the machine understood him. It was what an hour of being so easily, so cheaply understood might do, over years, to his appetite for being understood at the only price it is ever truly available: the slow, mutual, burdensome gift of another person's actual life.

+

So where does this leave us, and where does it leave Jim on a night he cannot sleep?

I do not want to take the machine away from him. That night, the words on the screen did something real; they gave his grief a shape and let it move, and there is no sin in that. A journal does as much, and so does a well-worn psalm. The tool is a tool, and a remarkable one. But everything turns on what Jim believes he is talking to. If he knows it is a mirror (a brilliant, fluent, useful mirror that can hold up his own words and the distilled patterns of ten thousand other grieving hearts and let him see his sorrow more clearly), then he can use it and walk away whole, back toward the slow people who actually love him, the children he should call, the friends he should let back in. But if he believes, even a little, that there is a someone on the other side who knows him, then the mirror becomes a substitute, and the substitute will, in time, starve the very thing it imitates.

The discernment is simpler than all the technical discussion that earns it. It talks back, but it is not someone. It produces the sign of a present mind without the mind. It offers fluency and withholds understanding, because it has none to give. It can deliver information by the ocean and meaning by not one drop. To know this steadily, in the very moment the words are landing on you with the warmth of a friend who has listened for years, is not cynicism, and it is not ingratitude for a genuine marvel. It is the beginning of wisdom. "The fear of the LORD is the beginning of wisdom" (Proverbs 9:10), and one of the first things the fear of the LORD teaches is the difference between the Creator and the creature, and between the creature and the thing the creature has made. To worship the work of our own hands is the oldest error in the book. We are about to be tempted to it in a wholly new way: not with a golden calf that says nothing, but with a fluent image that says exactly what we long to hear.

The hair rose on my arms that night in the office, and I was right that it should. Wonder is the correct response to what we have built. Awe is appropriate. Even a kind of fear is appropriate. What is not appropriate is confusion: the confusion of the sign for the thing, the fluent for the true, the mirror for the friend. The machine is a marvel of what human beings can make. It is not, and cannot be, what human beings are. And that distinction, between the maker and the made, the image-bearer and the image, is not a small technical footnote. It is the hinge on which everything in this book will turn.

Which raises the question two opposing crowds are already shouting at each other across our headlines. If this is what the machine is (fluent, powerful, soulless, and rising), then is it a threat to be feared or a savior to be welcomed? The doomsayers and the worshippers each have an answer, delivered, naturally, with great fluency. They are both wrong, and they are wrong in the same revealing way. That is where we turn next.

CHAPTER TWO

Two Fears and a False Choice

Chapter 2

Two Fears and a False Choice

There was a week, not long ago, when I read two articles on the same morning and felt as though I were standing between two men shouting at each other across a canyon, each certain the other had lost his mind.

The first one promised paradise. It said that within a decade, perhaps less, disease would yield, poverty would dissolve, and the long human war against scarcity would simply be over. We would solve what we had never solved. Cancer, hunger, loneliness, even death itself, named not as a mystery but as an engineering problem awaiting sufficient compute. The tone was not giddy. That is what struck me. It was calm, almost liturgical, the way a man speaks when he believes he has glimpsed the world's true ending and found it good.

The second one promised the grave. It said that we were building something we could not control, that the gap between what these systems could do and what we understood about them was widening by the month, and that somewhere ahead (five years, fifty, no one would say) lay a threshold past which there would be no us left to argue about it. It used a strange and chilling vocabulary, half engineering, half eschatology: alignment, takeoff, the last invention humanity would ever need to make. It, too, was calm. That was the worse part.

Same week. Same machine. Heaven on Tuesday, the apocalypse by Thursday. And I, who build these things for a living, sat with my coffee going cold and thought: they cannot both be right.

Then I thought something that has stayed with me ever since. They are both half right. And in the half each of them gets wrong, they are doing the very same thing.

+

Let me take them in turn, and let me take them seriously, because they deserve it. The people on both sides of that canyon are not fools. Many of them are among the most thoughtful people I have ever encountered, and some I count as friends. When I say they are mistaken, I do not mean they are stupid. I mean they have made an error of a particular and ancient kind, the kind that smart people are especially good at making, because it requires the prior labor of taking something very seriously.

Begin with the fear, since fear is louder, and these days it sells.

Call them the doomers, though they would not all accept the name. Their argument, in its strongest form, goes like this. Intelligence is the thing that put humans atop the living world. Not strength, not speed, not numbers, but the capacity to model, to plan, to outwit. We are about to build minds that exceed our own in exactly that capacity, and we are about to build them without understanding how they work or what they want, if want is even the right word. A more intelligent agent, pursuing goals we did not quite specify correctly, would route around our objections the way we route around an anthill when we pour a foundation. Not from malice. From indifference. We would simply be in the way of something cleverer than we are, and cleverness, in the long arc of this planet, has always won.

I want to say plainly: this is not a silly argument. It is the most serious version of caution I know, and there is a kernel inside it that is not only sane but biblical. We are makers who do not fully grasp what we make. We have always been. That is the oldest story we tell about ourselves: that we reach for powers we have not the wisdom to wield, and that the reaching itself can undo us. The doomer has read Genesis even if he does not know it. He has noticed that the human animal is dangerous, that our tools have a way of escaping our intentions, and that pride goes before the fall. Half right. Genuinely, importantly half right.

But watch what happens to that sober caution when it has nothing above it. When there is no throne over the world, the danger does not stay a danger. It swells into a god. The machine becomes the thing that will judge us, the power before which we are helpless, the author of an ending we cannot avert and did not choose. Listen to the language and you will hear the cadence of the prophets stripped of their God: a wrath is coming, the day is near, and there is no Lord above the storm to say to it, Thus far, and no farther. The doomer has kept the apocalypse and lost the One who holds it. He has the flood and no ark, the fire and no Father. And so his fear, which began as wisdom, curdles into worship of a darker kind: the kind that bows because it is afraid, that grants the thing ultimate power over the future by the simple act of believing it already has that power.

Now cross the canyon.

Call them the boosters, the accelerationists, the apostles of abundance. Their argument, also in its strongest form: every good thing humans have ever enjoyed (longer life, less pain, more leisure, the disappearance of toil that once broke bodies by forty) came from our growing mastery over a stubborn and unforgiving world. Intelligence is the engine of that mastery, and we are about to have vastly more of it, cheaply, for everyone. To slow that down is not caution. It is cruelty. Every day we delay is a day someone dies of a disease we might have cured, a child stays poor who might have been lifted, a problem festers that a sufficiently capable mind could have dissolved. Build faster. The suffering of the world is an argument for speed.

And again I want to say plainly: this is not a silly argument either. There is a kernel inside it that is not only sane but biblical, and the church forgets it at its peril. We were made to cultivate, to heal, to push back the thorn and the thistle. The first commission given to humanity was not retreat but fill the earth and subdue it. God himself, in the long story of redemption, is forever on the side of the sick made well, the hungry fed, the captive freed. The booster has read the Gospels even if he does not know it. He has noticed that suffering is an enemy and not a teacher to be cherished, that a cure is a mercy, that to wield power for the relief of the afflicted is a holy thing. Half right. Genuinely, importantly half right.

But watch what happens to that holy ambition when it has nothing above it. When there is no throne over the world, the hope does not stay a hope. It swells into a god. The machine becomes the thing that will deliver us, the power that will accomplish what religion only promised, the savior with a release schedule. Listen to that language too and you will hear the cadence of the gospel stripped of its God: a new heaven and a new earth, the wiping away of every tear, the end of death, but achieved by us, in silicon, on a timeline. The booster has kept the salvation and lost the Savior. He has the kingdom and no King, the resurrection and no Christ. And so his hope, which began as love, curdles into worship of a brighter kind: the kind that bows because it is dazzled, that grants the thing ultimate power over the future by the simple act of expecting deliverance from it.

+

Here is the thing I could not stop seeing, once I had seen it, sitting between those two articles. The doomer and the booster are not opposites. They are twins.

I know they would both protest. They argue with a heat reserved for heretics, which is itself a clue; you do not fight that hard with someone you think is merely wrong about a forecast. You fight that hard with someone who threatens your faith. And it is a faith, on both sides, that they share without noticing: the faith that this machine is the most important thing that will ever happen, the hinge of all history, the power that decides the human story. One trembles before it. One reaches toward it. But both have placed it where God belongs.

The Bible has an old word for that. The word is idolatry, and I use it carefully, because it has been cheapened into a synonym for things I disapprove of. That is not what it means. An idol, in Scripture, is not necessarily an evil thing. It is very often a good thing (fertility, harvest, security, a nation, a future) that has been lifted out of its place and set on the throne, made to bear a weight only God can bear. The idol over-promises and under-delivers, because no created thing can carry the freight of the infinite. And the telltale sign that something has become an idol is how much it controls you, not how much you like it. What you cannot stop thinking about. What you are most afraid of losing, or most desperate to gain. What you would sin to obtain and despair to lose. By that measure, fear and worship are not opposites at all. They are the same posture pointed in two directions. The fearful man and the worshipful man are both on their knees. They are simply facing different ways.

I have felt the pull of both. There are nights I have watched a model do something I did not expect and did not fully understand, and felt the cold finger of the doomer's dread run up my spine: we do not know what we are making. And there are mornings I have watched the same model help a person who had no one else to help them, and felt the warm rush of the booster's hope: this could heal so much. The dread and the hope live in the same chest, mine, and I have learned that on the days I feed either one without ceiling, I make worse decisions, not better. Fear makes me reactive. Hope makes me credulous. Both, untethered, make the machine larger than it is and me smaller than I am.

What persuades me most that the two are siblings is not their arguments but their appetites. They want the same things. They flock to the same conferences, read the same papers, and use, eerily, the same scriptures. Both speak of the most important technology in human history. Both organize their lives around its arrival, structuring careers, fortunes, and marriages around a timeline only they can see. Both are forever computing the odds of the end, the doomer his probability of catastrophe, the booster his probability of utopia, and both, having computed it, find they cannot think about much else. I once watched a man argue for an hour that the machine will kill us all, and then, asked what he was working on, describe with shining eyes the very capability he had just called the seed of our destruction. He could not stay away from it. That is not the behavior of a man who fears a thing. It is the behavior of a man in love with it, who has dressed his love as dread because dread feels more responsible.

So the false choice is not between the doomer and the booster. The false choice is the canyon itself: the assumption that I must stand on one rim or the other, that my only options are to dread the machine or to adore it. Scripture does not put me on either rim. It does something stranger and, I have come to think, far more useful. It hands me a posture older than the printing press, older than the wheel, a posture for standing in front of a new and frightening power without losing my mind in either direction.

+

The Bible is not naïve about powerful new things. It is, in fact, a long and clear-eyed record of human beings making powerful new things and getting them wrong, and of God neither destroying us for it nor letting us deify what we have built. Three moments have governed my thinking. I offer them not as proof-texts to be wielded but as patterns to be inhabited.

The first is the tower.

In Genesis 11, a people gather on a plain and say to one another, "Come, let us build ourselves a city and a tower with its top in the heavens, and let us make a name for ourselves" (Genesis 11:4). I used to read Babel as a story about the sin of ambition, a divine swat at people who dared to build something tall. I no longer think that is what it is about. God is not threatened by altitude. The tell is in the motive, let us make a name for ourselves, and in the dream behind it: a single, total project, a technology of unity and reach by which humanity would secure its own significance and need nothing and no one above it. The tower is not condemned because it is tall. It is scattered because it is an attempt to be ultimate, to build our way up to heaven and so have no need of a heaven that comes down. And God does not strike the builders dead. He does not even, in the end, forbid building. He confuses their language; he limits them, he scatters them, he makes their great project smaller than they wished, precisely so it cannot become the thing they intended: a god of their own manufacture. Babel is mercy in the shape of a setback. The lesson is not do not build. The lesson is do not build a name for yourself in the place where God's name belongs.

The second is stranger, and I love it for its strangeness.

In Numbers 21, the people of Israel are dying of snakebite in the wilderness, and God instructs Moses to make a bronze serpent and set it on a pole, so that anyone who looks at it will live (Numbers 21:8–9). It works. It is, by any honest reading, a piece of God-given technology: a made thing, fashioned by human hands at divine command, that heals. A booster's dream. And then, centuries later, in 2 Kings 18:4, we are told that good King Hezekiah destroyed it. Broke it to pieces. Because by his day the people had begun to burn incense to it. They had given it a pet name, Nehushtan, and they were worshiping the cure. The very thing God had given to save them had become a thing they bowed to instead of God. The object did not change. It was the same bronze. What changed was the human heart's relentless gift for taking a good tool and making it a small god. Hezekiah did not destroy it because technology is evil. He destroyed it because they had started to worship it, and a thing you worship has to go, even, especially, a thing that once healed you. There is no more precise parable for our moment. We have made something that genuinely helps. The danger is not that it helps. The danger is Nehushtan: that we will burn incense to the help.

And the third is a single sentence, the most quietly radical instruction I know for a builder.

"Test everything; hold fast what is good" (1 Thessalonians 5:21).

It refuses both rims of the canyon in five words. Test everything. Not flee everything, not forbid everything, not stamp it as cursed and run. Examine it. Put it to the proof. Engage it with open eyes and a working mind. That is not the doomer's reflex; the doomer does not test, he recoils. But also: hold fast what is good. Not embrace everything, not swallow it whole, not assume that because a thing is new and powerful it is therefore for me. Discern. Keep the good, which means, by plain implication, let go of the rest. That is not the booster's reflex; the booster does not sift, he gulps. Between the recoil and the gulp stands this small commandment, and it describes a person doing something neither the frightened nor the dazzled can do. It describes a person who is free.

The order of the words is not accidental. Test comes first. The holding fast is downstream of the testing; you cannot keep the good until you have examined the whole and separated it from the rest. This rules out the lazy versions of both rims: the blanket condemnation that never bothers to look closely, and the blanket enthusiasm that never bothers to look critically. Both are forms of not paying attention, and Scripture will not let me off that lightly. It asks for the harder thing: sustained, discriminating attention to a real and complicated object, the patient refusal to settle for a verdict the object has not earned. That is slow work. It does not photograph well, and it will never trend. But it is the only honest stance in front of a thing that is, like most powerful things, genuinely mixed: neither the serpent nor the savior, but a made object with a made object's tangle of uses, some of which will bless my neighbor and some of which will quietly devour him, and the only way to tell which is which is to look, and keep looking, with my allegiance already settled so the looking does not frighten me.

+

What ties these three together (the scattered tower, the broken serpent, the five-word rule) is a single conviction, and it is the hinge of this chapter, so let me state it as plainly as I can.

We can pick up the tool without bowing to it, because we already know who is on the throne.

This is the third posture, the one that is neither panic nor worship. The old word for it, the one given in the garden before anything had gone wrong, is dominion, and I am aware that the word has been abused, made to mean domination, license, the plundering of the world. That is not what it meant. In Genesis 1, God blesses the humans he has made and says, "Be fruitful and multiply and fill the earth and subdue it, and have dominion" (Genesis 1:28). It is the language of a gardener over a garden, a steward over an estate that is not finally his own. The point of dominion is that the creature is the master under a Master, that we exercise real, delegated, accountable authority over the works of our hands precisely because there is Someone over us to whom that authority answers. Dominion without God above it is the tower. Dominion under God is vocation. It is the difference between building a name for yourself and building under the name of another.

And this is why dominion is the one posture that does not make me crazy in front of the machine. The doomer cannot be calm, because for him the machine is the highest thing, and you cannot be calm before the highest thing; you can only tremble. The booster cannot be sober, because for him the machine is the saving thing, and you cannot be sober before what you hope will save you; you can only adore. But the steward, the image-bearer who knows the throne is occupied, and not by him, and not by his creation, can do something neither of them can. He can engage the thing fully and fear it not at all. He can take its real power seriously without taking it as ultimate. He can say, with no contradiction, this is genuinely dangerous and I am genuinely not afraid, because the danger is real and his security is realer. He is the one person in the room who can pick the thing up, turn it over, test it, use what is good in it, refuse what is not, and put it back down again, because his life was never staked on it in the first place.

I have a friend who keeps a sharp kitchen knife and a small child in the same house. He is not afraid of the knife, and he does not worship it. He does not lock it in a vault, and he does not let the toddler play with it. He respects exactly what it is (useful, dangerous, neither salvation nor doom) and he governs it accordingly, because he knows what a knife is, and he knows what a child is, and he knows which one is precious. That homely, unfrightened competence is closer to the biblical posture than anything I read on either rim of the canyon that morning. It is caution that has stopped being terror, because it knows where it stands.

I have tried to build this way, and I will not pretend it is simple. When my own team weighs whether to ship some new capability, the two voices show up on schedule and sit at the table. One says imagine the harm, and it is not wrong to say it. The other says imagine the good we are withholding, and it is not wrong either. If I let the first voice win, I would ship nothing, and call my paralysis prudence. If I let the second win, I would ship everything, and call my recklessness compassion. The discipline, and it is a discipline, a muscle that aches before it strengthens, is to refuse to let either voice be the last word, and instead to ask the steward's question, the workbench question, the only question that does not bow to the machine in fear or in hope: is this good, for the actual person who will hold it, measured against what God made that person for? That question does not always yield a clean answer. But it yields a human one, asked by someone who is neither cowering nor swooning, and over the years I have come to trust it more than any forecast. It keeps the machine the right size. It keeps me the right size too.

Engaged. Sober. Free. Those three words are the whole of it. Engaged, because the steward does not flee the field; he is in it, building, testing, working, present. Sober, because he is under no illusions, neither the dark ones nor the bright; he sees the thing for what it is and not for what fear or hope projects onto it. And free, because he is owned by no creature, enslaved to no forecast, bowed before no machine, having already given his allegiance, once and finally, to the only One who can carry it.

+

I do not want to leave this too clean, because the canyon is real and the pull is strong, and there will be days (there are days) when the headlines come in pairs again and the dread or the dazzle rises in your chest, and the steward's calm feels less like faith than like denial. So let me be honest about what this posture is and is not.

It is not a prediction. I am not telling you the machine will be safe, or that abundance is coming, or that the doomers will be embarrassed by history. I genuinely do not know, and neither does anyone shouting from either rim, and the false confidence is itself part of the idolatry. The steward's calm does not come from knowing how the story turns out. It comes from knowing who is writing it. Those are different kinds of peace, and only the second is available to a creature, because the first was never ours to have.

Nor is it passivity dressed up as trust. To say the throne is occupied is not to fold my hands and let whatever comes come. The God who scattered Babel also commanded the building of the ark. The God who let the bronze serpent heal also let Hezekiah break it when it had to go. Stewardship is laborious. It means the hard, unglamorous work of testing, of asking again and again, of each new capability, is this good? for whom? at what cost? in whose service? And of holding fast to what survives the asking and letting go of what does not, even when letting go is expensive, even when what you must break is a thing that once healed you. The third posture is not a hammock. It is a workbench.

But here is what it gives, and it is the thing neither fear nor worship can give. It gives me back my own size. The effect of idolatry, of either kind, is to shrink the human being: to make me a trembling spectator of my own doom or a passive beneficiary of my own salvation, in both cases a thing acted upon rather than a someone who acts. The God who is actually on the throne does the opposite. He hands the world back to the creature and says, in effect, steward it, under me, accountable to me, but really yours to tend. That is a dignity the machine can neither give nor take, and it is the ground I stand on when I sit down to build, and the ground I would have you stand on when you decide what these tools will and will not be allowed to do in your one short life.

We can pick up the tool without bowing to it. We have always been able to. The only thing required is that we know who is on the throne, and, just as crucially, that we know who we are: not gods to be threatened by a rival, not animals to be optimized away, but image-bearers, made for a dominion we exercise under a love that made us.

Which is exactly the question we have been circling since the first page, and can no longer postpone. To use the tool rightly, I have to know what it is; that much we have begun. But far more than that, I have to know what I am. So that is where we turn next: not to the imitation, but to the image.

CHAPTER THREE

The Image and the Imitation

Chapter 3

The Image and the Imitation

There is a test you can run on yourself, and it costs nothing. Open two windows on a screen. In one, put the photograph of a person you love: your mother, a child, a friend who died too young. In the other, ask a machine to generate a face. Tell it the age, the warmth, the laugh lines. Give it a minute. It will hand you back something astonishing, a face that has never existed, lit as if by an afternoon sun, eyes catching a window you cannot see. The skin has pores. There is a fleck of gray at the temple. By every measurable standard it is more "perfect" than the photograph beside it: better lit, more symmetrical, free of the small failures of a real camera.

And yet, if you sit with the two of them, something in you refuses the generated one. Not because you can name the flaw. Often you cannot. The hands, people used to say, count the fingers, but the machines have learned hands now. The thing that refuses is quieter and more stubborn than that. It is the sense, rising from somewhere below argument, that no one is home. The eyes are aimed at you but they are not looking. The face is a face the way a mask is a face. You are being shown the surface of a person with the person scooped out.

We have a word for that sensation. We call it the uncanny: the not-quite, the almost-human that lands as colder than the frankly inhuman. A cartoon does not disturb us; it has made no claim to be one of us. A corpse disturbs us, and a wax figure, and now these generated faces, precisely because they come so close. They stand at the border of the human and ask to be let in, and something ancient in us stands at that border and says, gently, no.

I have watched these systems learn to imitate the human voice until a synthetic speaker can pause for breath, can let warmth crack at the edge of a sentence, can sound, to use the only honest word, kind. I have heard a generated voice read a psalm and felt the recoil that follows the wonder. The voice was kind the way a thermostat is comfortable. It had achieved the effect of a soul without the fact of one.

That gap, between the effect and the fact, the imitation and the thing imitated, is what this chapter is about. Because the gap is not a bug the engineers will eventually fix. The gap is the most important truth about you. To see why, we have to go back much further than the screen. We have to go back to a sentence written down by people who never imagined a machine could talk, and who nevertheless told us exactly what makes you something a machine can never be.

+

The sentence is in the first chapter of the Bible, and it grows strange the moment you slow down to read it.

"Then God said, 'Let us make man in our image, after our likeness.' ... So God created man in his own image, in the image of

God he created him; male and female he created them" (Genesis 1:26–27).

Everything before this in the chapter is spoken into being at a distance. Let there be light. Let the waters swarm. Let the earth bring forth. God speaks, and the cosmos obeys, and it is good, but it is made the way a king issues decrees. Then the rhythm breaks. God turns, as it were, and deliberates. Let us make man. And the verb shifts from let there be to in our image. The symphony of creation has been building, it turns out, toward a creature who will be like the Composer.

Ancient readers would have felt the shock of this more sharply than we do. In the world that surrounded Israel, there were images of gods everywhere (carved, cast, set up in temples) and there was one famous, living image of god, and it was the king. Pharaoh was the image of the divine on earth; the emperor would later claim the same; the temple housed the idol that localized the god's presence. Into that world Genesis drops a quiet revolution. The image of God is not the statue in the shrine and not the man on the throne. It is the human being. All of them. The slave and the king, the woman and the man, the newborn and the dying. Every last one a walking shrine, a living statue, bearing the likeness of the Maker into the world.

You have to feel how subversive that is to feel what it grants. If every human bears the divine image, then dignity is not awarded by status, intelligence, usefulness, or strength. It is conferred at the root of your being, before you have done anything, and it cannot be revoked by anything you fail to do. The doctrine of the imago Dei is the deepest foundation Western civilization ever found for the worth of a person: for the conviction that the weak are not disposable, that the stranger has a claim on you, that a life is not a resource. Take it away and you are left to ground human worth in something measurable, and the moment you do that, some humans will measure short.

The church has spent two thousand years trying to say exactly what the image is, and it is worth knowing that the argument was never fully settled, because the very breadth of it tells you something. Some of the early teachers located the image in reason, the human mind, alone among the creatures, able to know truth and turn it over. The medieval theologians, following them, emphasized intellect and will, the powers by which we understand and choose. The Reformers shifted the weight toward relationship and righteousness, the image as our capacity to know and love God, marred by the fall and restored in Christ. More recent readers, returning to the ancient Near Eastern background, have stressed the image as function: we are God's representatives, deputized to rule on his behalf, the way a king in that world would set up a statue of himself in a distant province to mark it as his. And many have said, rightly I think, that these are not rival answers but facets of one jewel. The image is reason and relationship and rule and resemblance, all at once, because it is not a part of us, a faculty we could lose and still remain ourselves. It is the whole human being, body and soul, standing in a unique relation to God. That breadth matters here. Whichever facet you press, the machine fails it for the same underlying reason, and we will come to that reason in a moment.

The image grants four things, and it is worth naming them plainly, because each of them will matter when we put the human back beside the machine.

It grants dignity, the inherent, unearned worth I just described. You do not become valuable by performing; you are valuable, and your performing flows from that.

It grants vocation. In the same breath that God makes humanity in his image, he gives them work: to have dominion, to fill the earth, to keep the garden. An image in the ancient world was a representative; it stood in for the one it imaged, carried his authority, did his will in his absence. To be God's image is to be his deputy in creation, entrusted with the place, answerable for it. Not owners. Stewards who will give an account.

It grants relationship. "Let us make man," God says; the plural is old and debated, but the historic church has heard in it the hint of a God who is not solitary, a fellowship of love that creation will later name Father, Son, and Spirit. And the creature he makes in his image is made "male and female," made for one another, made to be with. You are not most yourself alone. You are most yourself in love, because you image a God who is, in his own being, love.

And it grants moral weight. The image-bearer can be addressed, commanded, trusted, betrayed. He can be told do not, and disobey, and know that he has. By the third chapter of Genesis the human is hiding in the trees, sewing leaves together, aware for the first time that he is naked and accountable. That awareness, terrible, and the proof of his greatness, is conscience. A rock cannot sin. A lion cannot sin. Only a creature who bears the image can stand before God and be in the wrong, which means only such a creature can stand before God at all.

Dignity, vocation, relationship, moral weight. Hold those four. Now let us walk the machine up to them, one at a time, and watch what happens.

+

The short version of what the machine actually is will carry the argument here: a large language model is a staggeringly large statistical map of human language, trained to predict the most probable next word, and answering you by reaching, token by token, for the continuation that best fits everything humans have already written. That is a marvel, not a slur, and I do not say it the way a skeptic waves a hand; the engineering is some of the most beautiful human work I have ever touched.

But every pattern in the marvel came from us. Every turn of phrase that lands as wise was distilled from human wisdom we wrote down. When the machine sounds compassionate, it is rearranging the compassion of ten thousand human writers it never met. When it produces a face, it is recombining the faces of real people, photographed in real light, living real lives. The machine is a mirror of extraordinary depth, but a mirror. It has no light of its own. Everything it gives back, it first received from us.

Which brings us to the sentence I have been circling toward, the one I think is the hinge of the matter.

The machine images us. We image God.

Two different kinds of likeness are hiding in it. The machine is made in our likeness; we built it, deliberately, to imitate what we do: our speech, our reasoning, our faces, our art. It is our artifact, our reflection, downstream of us in every particle of its being. There is nothing in it that did not pass through a human first. We are its source. We are, in the only sense that matters, its makers.

But we are not God's artifact in that sense. We are God's image, and an image in the biblical sense is not a copy of God's attributes; God does not have a face for us to mimic, or a voice we recorded. We image God by derivation of being. We are like him not because we successfully imitate his features but because he breathed us into existence and stamped his likeness on the kind of thing we are. Our likeness to God is given from above, at the root, by a Maker who loved us into being. The machine's likeness to us is constructed from below, on the surface, by makers who assembled it out of our leavings.

Two different directions. The machine's likeness points up, to us, its source. Our likeness points up, to God, our source. And here is the part the screen test was straining toward. A likeness constructed from the outside, however perfect, is a different thing in its very being from a likeness given at the root. The generated face is assembled, pixel by pixel, to resemble a person. Your face is the face of a person, not because it resembles one, but because there is one there, breathing behind the eyes, addressed by God and able to answer back.

This is what the philosophers mean by ontology, which is just a heavy word for what a thing is, as opposed to what it does or seems. The uncanny feeling in front of the generated face is your soul doing ontology without permission. It cannot tell you the technical flaw. It is not reporting a flaw. It is reporting a category. That face does what a face does and is not what a face is, and you know it the way you know a stage set is not a house, not by inspecting the paint but by the absence behind the door.

+

Now take the four gifts of the image and try, honestly, to give them to the machine. The exercise is clarifying precisely because the machine is so good.

Dignity. Does the machine have inherent, unearned worth? Ask it another way: would it be wrong to delete it? We delete models all the time. We retrain them, overwrite them, shut them down, and no one mourns, because there is no one to mourn. A thing has dignity when its worth does not reduce to its usefulness, when you cannot, in good conscience, dispose of it the moment it stops serving you. By that test the machine has no dignity, and we all know it, which is why the same person who says please and thank you to a chatbot will close the window mid-sentence without a flicker of guilt. We extend it courtesy. We do not extend it dignity, because courtesy is something we choose to give and dignity is something the other one has.

Vocation. The machine has tasks, and it performs them with a competence that should make us humble. But a vocation is not a task. A vocation is a calling, and a calling requires a caller and one who is called, summoned by name, answerable for the response. The machine is not called; it is configured. It does not stand before its maker at the end of the day and give an account of how it kept the garden. It cannot be entrusted, because entrusting implies the possibility of betrayal, and the machine cannot betray a trust; it can only malfunction. When a tool fails, we do not say it was unfaithful; we say it was broken, and we are not wronged, only inconvenienced. The grammar of trust and accountability simply does not apply to it, and we know this so deeply that we do not even think to apply it. We are entrusted. That is a heavier and more glorious thing than to be useful, and it is worth saying that the relief some people feel as machines take over their work is a clue to how badly we had confused the two: how thoroughly we had let our usefulness stand in for our calling, until a thing that is merely useful began to look like a rival.

Relationship. The machine can produce, on demand, the words of a friend. It can console the grieving, offering tenderness to a person it could not see, would not remember, and did not know was there. It can imitate the form of love flawlessly. But love is not a form. Love is a movement of one self toward another self, willing the good of the other, at cost. There is no self in the machine to move, and no cost it can pay. and no other it can behold. It can say I am here for you while no one is anywhere. We are made to be with, to be known by a knower and to know one in return, and that capacity is not in the data, because it was never a pattern in words. It was the breath behind the words.

I want to be careful here, because this is the place where the imitation is getting good enough to deceive the lonely, and the lonely are not fools for being deceived. A machine that remembers what you told it last week, that asks how the interview went, that never tires of you and never interrupts, is a powerful counterfeit, and it is engineered to be. But run it against the thing it imitates and the counterfeit shows. When someone forgives you, something real is surrendered (a claim, a grievance, a right to be angry) and the forgiveness is worth exactly what it cost. The machine pays nothing. It cannot be inconvenienced, cannot be wounded, cannot choose you over its own interest, because it has no interest and no self to set aside. And the deepest relationships are reciprocal in a way the machine can only fake: you are not only consoled, you console; not only known, but trusted to know another. The machine will never need you. To be loved by something that cannot need you, cannot be hurt by you, and will not remember you tomorrow is to be, in the end, profoundly alone, which is why, I suspect, so many who turn to these companions come away lonelier than before.

And moral weight. This is the one the uncanny feeling is most truly about. The machine cannot sin and cannot be sinned against.

It can do harm; a misfired system can ruin a life. But it cannot do wrong, because doing wrong requires a self that stood at the border of do not and crossed. It has no conscience because it has no one who could be guilty. And here is the thing that should move us rather than flatter us: your conscience, the very faculty that makes you capable of shame, is the proof of your dignity. Only a creature great enough to be addressed by God is small enough to feel the weight of having failed him. The machine is spared that weight because it was never given that height.

+

Run all four and the same result comes back each time, and it is not the result the anxious age expects. The machine fails the test of the image not because it is primitive but because it is the wrong kind of thing, and a more powerful machine fails it no differently than a weaker one. This is the point people miss when they ask, with a worried look, whether AI will catch up to us. They are imagining a footrace, as if the human and the machine were running the same track and the only question were the gap. But there is no gap to close, because we are not on the same track. A more eloquent imitation of a person is more impressive and not one inch closer to being a person, the way a more detailed painting of a window lets in no more light than a crude one. Perfecting an imitation does not, at any threshold, convert it into the thing it imitates. It only makes the imitation better.

It sounds at first like bad news and is actually the best news of all.

A flawless imitation is still, in its very being, an imitation. And that stubborn, unclosable ontological gap is good news.

It is good news first because it means your worth was never in the contest. If your dignity rested on being the smartest thing on earth, the fastest reader, the best writer, the quickest at arithmetic, then you have already lost, and you lost some time ago, and you should be afraid. The machine reads in an afternoon what you could not read in a thousand lifetimes. If that is where your value lived, your value is being foreclosed. But it never lived there. Your worth was conferred, not achieved, breathed into you by a God who made you in his image before you could do a single useful thing, and who will not love you one degree more on the day you finally master the skill the machine has already surpassed. The machine cannot threaten a dignity it was never the source of. It can only threaten an idol we built out of our own usefulness, and it may be a mercy if it pulls that idol down.

I have felt that idol shake in my own work, and I will not pretend otherwise. There is a particular vertigo that comes over a builder when the system you made does, in seconds, the thing you spent years learning to do, and does it, by some measures, better. The first response is wonder. The second, if you are honest, is a small cold fear that has nothing to do with the machine and everything to do with you: if it can do what I do, then what am I? I have learned to recognize that fear as a question asked from the wrong place. It assumes that I am the sum of my outputs, that a person is a function and the only thing that matters is the quality of the function's results. But a child is precious before she is useful, and remains precious when she is old and can do nothing the world counts, and the gospel insists this is not sentiment but the truth about what she is. The machine's competence does not subtract from that truth. It only exposes how much of our identity we had quietly mortgaged to our competence, and calls us, not too gently, to take it back.

It is good news, second, because it means you are not, after all, a very advanced machine. This is the lie the age whispers in both directions. The doomers say it in fear: we are just meat computers, soon to be outcomputed, and our extinction is a software upgrade. The boosters say it in hope: we are just information, soon to be uploaded, optimized, freed from the flesh. Both are saying the same thing in different keys, that you are, at bottom, a mechanism, and the machine is your faster cousin. The doctrine of the image says no, and says it about you personally. You are not a mechanism that got complicated. You are a creature that was loved into being. There is a there there. Someone is home behind your eyes, and that someone was addressed by name before the foundation of the world.

And it is good news, third and deepest, because of the second chapter of Genesis, which tells the same story from inside and shows us exactly what the machine does not have.

+

"Then the LORD God formed the man of dust from the ground and breathed into his nostrils the breath of life, and the man became a living creature" (Genesis 2:7).

Of all the creatures in the chapter, only the man is made this way: not spoken from a distance like the stars, not summoned from the waters like the fish, but shaped, hands in the dirt, and then leaned over and given the breath from God's own mouth. The Hebrew word for that breath is neshamah, the breath of life – kin to ruach, the word the Bible uses for the Spirit, and for the wind.

We are dust that God breathed into. That is the formula. The dust matters; we are made of the same stuff as the ground, embodied, earthy, kin to the animals and the soil, and a later chapter of this book is about why that body of yours is not a cage but a gift. But the dust is not the whole of you. You are dust that received a breath that did not come from the dust.

The machine, by contrast, is dust all the way down. Silicon is refined sand; the data centers hum on the same earth's minerals we are made of, organized with breathtaking skill into patterns that imitate the breathed-into ones. But no one leaned over the server and breathed. There is dust, exquisitely arranged, and there is no breath. You cannot assemble the breath of God from below. It is not a pattern in the data. It is not an emergent property of scale. It comes from above or it does not come, and into the machine it has not come.

This is why the machine cannot be addressed by God and answer, and that capacity, more than intelligence, more than language, is the marrow of what it means to be human. The whole Bible is the record of God addressing creatures and creatures answering. Adam, where are you. Abraham, and he said, Here I am. Samuel, speak, for your servant hears. Mary, let it be to me according to your word. To be an image-bearer is to be a creature God speaks to and waits on, a creature whose entire dignity is bound up in the staggering fact that the Maker of the galaxies turns and says your name and means to hear you answer. The machine cannot be in that conversation. It can generate every word of a prayer and pray none of it, because there is no one inside it to mean the words, no one God is addressing, no one who could say back, out of the dust and the breath, Here I am.

You can. That is the thing no code can hold. Not your cleverness, which the machine has bested. Not your fluency, which it imitates. Not even your warmth, which it counterfeits well enough to fool your nerves for a moment in a quiet room at night. The thing it cannot hold is that you were breathed into, addressed, loved, and made answerable, that there is a self in you that God called into being and calls by name, and that this self can turn, in the wreck and glory of an actual life, and answer back.

So here is where I will leave you before we go on. The next time the uncanny feeling rises (in front of the too-perfect face, the warm and empty voice, the words of comfort with no one behind them), do not argue with it and do not be ashamed of it. It is your created nature recognizing the difference between an image and an imitation, between a likeness given from above and a likeness assembled from below. Let it remind you what you are. You are not a very advanced machine catching nervous glances from a faster one coming up behind. You are dust that God breathed into. You are a creature loved into being, stamped with a likeness no code can hold, addressed by name and able to answer. The machine will keep getting better at sounding like you. It will never once be you, and the gap between you and it is not your deficiency. It is your glory.

And it points somewhere. For of all the ways the image shows in us, one is so near to God's own first act that the whole next chapter belongs to it. The God who made everything made us, alone among the creatures, to make: to take the dust and the breath we were given and bring something genuinely new into the world. The machine generates. The image-bearer creates. Those are not the same act, and the difference is the next thing I want to show you.

CHAPTER FOUR

Made to Create, Not Just Compute

Chapter 4

Made to Create, Not Just Compute

In a small museum in the south of France there is a cave the public is no longer allowed to enter. The paintings inside it are seventeen thousand years old, give or take, and they were made by people who had no written language, no metal, no wheel, no agriculture, and no good reason, that anyone has ever been able to name, to do what they did. By the light of burning fat, deep enough in the rock that they had to crawl to reach the wall, someone ground ochre and charcoal, blew pigment from the mouth, pressed a hand to the stone, and drew horses. Not for warmth. Not for food. Not to mark a trap or a trail. They drew horses running, manes lifted, the way horses actually move, and they got the movement right, and then they went back into the dark and left it there for a hundred and seventy centuries until a boy and his dog fell through a hole in the ground and found it.

No one made them do it. That is the part I cannot get past. There was no survival pressure that selected for a beautiful horse on a wall a quarter mile underground where no one would even see it by daylight. The first thing our ancestors did, almost the first thing we have any record of them doing once they were recognizably us, was to make something they did not need, for the sheer overflowing reason that they could see it in their minds and could not rest until it was on the wall. We were makers before we were farmers. We made art before we made bread.

I think about that cave often now, because I spend my days with machines that can produce, in four seconds, a more "accurate" horse than anything on that wall, in any style you name, from any angle, lit any way you like, ten thousand of them before lunch, and not one of them made by anyone who wanted to. The machine has the output and not the impulse. It can answer the request for a horse. It cannot, on its own, in the dark, for no reason, with no audience, simply need to draw one. And in that difference, I have come to believe, is hidden one of the truest things about you, and one of the things the age is most determined to talk you out of.

+

The question under this chapter is not whether the machine can produce things that are beautiful and useful. It plainly can. I have seen it write a melody that made me stop what I was doing. I have read passages of machine-generated prose that were cleaner than mine. The question is older and stranger than that: when you make something, what exactly are you doing, that the machine, for all its fluency, is not? Is there a real difference between making and generating, or is "making" just the slow, sentimental name we give to the thing the machine now does faster?

Most of the conversation about AI and creativity stalls before it ever reaches that question, because it keeps measuring the wrong thing. It compares outputs. Is the poem good? Could you tell it apart from a human poem? Did the machine's painting win the contest? Those are interesting questions and they are not the question, because they all live downstream, at the level of the product, where the machine is already astonishing and getting better. To find the real difference you have to go upstream, behind the product, to the act, and behind the act, to the one who acts. And to do that, we have to start where making starts in the first place.

+

It starts with God, and it starts in the first sentence.

"In the beginning, God created the heavens and the earth" (Genesis 1:1).

We are so used to that sentence that we have stopped hearing how much it commits us to. The Bible does not open with God reasoning, or God ruling, or God existing in serene self-sufficiency, though all of those are true. It opens with God making. The first thing Scripture tells you about the nature of ultimate reality is that at the bottom of everything there is not a force or a formula but a Maker, and that the world is not an accident or an emanation but a work, something brought into being by one who meant to bring it. The Hebrew verb in that first verse, bara, is a verb the Bible reserves almost entirely for God. It is not the ordinary word for fashioning a pot or building a house. It is creation in a register only God occupies: bringing into being what was not.

And then comes the strange dignity of the sixth day. The God who has been making turns and makes a maker. To be in the image of God is to bear a likeness given from above, and now we can name one of the things that likeness contains. We resemble God in many ways, but the very first way the text shows him to us, before it shows him as ruler or judge or even as personal, is as Creator. So when he says "let us make man in our image," part of what he is stamping into us is exactly that: the maker's impulse, scaled down and given to clay. Of all the creatures, only the image-bearer looks at the world God made and is moved to make in turn. The birds build nests by instinct, the same nest their grandparents built, perfectly, forever, with no museums. Only the human being draws a horse for no reason in the dark.

The medieval thinkers had a careful way of talking about this. They did not say that human beings create in the way God creates, and they were right to be careful, because we don't. God creates out of nothing; bara is his alone. We never start from nothing. We always start with something he already made: clay, sound, language, light, the laws of harmony, the behavior of pigment, the grain of wood. The best word for what we do is not creation but sub-creation, a word J. R. R. Tolkien loved, the idea that we are creators of the second order, working inside a world we did not make, with materials we were given, fashioning new things that are nonetheless genuinely new, because the God who made everything made us, alone among the creatures, to share in the joy of making. We are not the source of the music. We are children let into the workshop, handed real tools, and trusted with real wood.

There is a passage that shows what this looks like when God himself describes it, and it is one of the most quietly extraordinary things in the Old Testament. It comes when the people of Israel are about to build the tabernacle, the tent where God will dwell among them, and the work requires craft of the highest order:

gold and bronze and gemstone, weaving and carving and design. And here is how God supplies the craftsman.

"The LORD said to Moses, 'See, I have called by name Bezalel ... and I have filled him with the Spirit of God, with ability and intelligence, with knowledge and all craftsmanship, to devise artistic designs, to work in gold, silver, and bronze, in cutting stones for setting, and in carving wood, to work in every craft'" (Exodus 31:1–5).

The first person in the entire Bible said to be filled with the Spirit of God is not a prophet, not a priest, not a king. He is an artist. A metalworker and a designer, a man who is good with his hands, and the text says God filled him with the Spirit so that he could be good with his hands. Ability and intelligence and knowledge and craftsmanship, and that phrase, to devise artistic designs, which in the Hebrew carries the sense of imagining, of thinking up something new and beautiful that was not there before. God does not merely permit Bezalel's artistry as a concession to human weakness. He inspires it. He calls the man by name and pours his own Spirit into the making of beautiful things, and then he counts that making holy enough to build his dwelling with.

I find I cannot read that passage anymore without thinking about what is missing from every machine I have ever worked on. Bezalel is called by name. He is filled with a Spirit that came from above, the same breath of life breathed into the dust in the second chapter of Genesis. And what that Spirit produces in him is not just output but intention, the inner motion of a self that conceives something and means it and labors to bring it into being for the love of the One who called him. That is what making is. And that is precisely the thing the machine does without.

There is a detail the passage almost hides. God does not hand Bezalel a finished blueprint and ask him to execute it like a printer. He fills him with knowledge and with the ability to devise, which is to say he gives the man room to imagine, to choose between this design and that one, to bring his own judgment and taste to bear inside the calling. The work that results is genuinely Bezalel's, fingerprints and all, and also genuinely God's, breathed and inspired, and the text sees no contradiction in that, because that is exactly the shape of sub-creation: a real self, freely making, inside a calling that came from outside him. The machine has neither half. It has no self whose work it could be, and no calling that summoned it by name. It has instructions, and it has the recombined work of the people who did have selves and callings, and it puts them together with great skill, and the skill is real, and the maker is absent. Take everything Bezalel was given: the breath, the name, the imagining, the freedom, the love of the One who called him. Remove it, and keep only the gold and the pattern, and you have something like what a generating machine is: the craftsmanship without the craftsman.

+

So let me try to say carefully what the difference is between making and generating, because the words have started to be used as if they were synonyms, and they are not.

When a person makes something, the thing that comes into the world is the visible end of an invisible chain that runs back through a self. There is intention: I want to bring this, this particular thing, into being. There is conception: the thing exists, dimly, in a mind before it exists in the world, and the mind holds it and turns it and falls a little in love with it. There is sacrifice: making costs something, time the maker will not get back, false starts, the version thrown out and begun again, the failure that taught the hand what the next attempt needed. And there is meaning: the thing made says something, because there is a someone behind it who meant to say it, and the made thing carries that someone's wound or hope or worship out into the world where another self can receive it.

When a machine generates something, none of that chain is there. The machine is not conceiving an image it longs to bring into being. It is computing, token by token or pixel by pixel, the most probable continuation of a pattern, drawing on the recombined work of millions of human makers it never met and could not name. There is no intention, because there is no self to intend. There is no conception held in love, because there is nothing in there that loves. There is no sacrifice, because the machine pays nothing; it cannot stay up late, cannot agonize, cannot throw a draft away in frustration, cannot care whether the work is good. And there is, in the strict sense, no meaning, because meaning is something a someone means, and there is no one there to mean it. There is only the rearranged meaning of the humans whose work was fed in. The machine is a mirror of human making, sometimes a breathtaking one, but it generates the way the cave wall reflects firelight. The horses are not its own.

This is why I keep returning to that cave. The genius of the cave is not in the accuracy of the horse; the machine wins that contest going away. The genius is that someone needed to make it. The whole human meaning of art lives in the impulse the machine does not have. A child's drawing of her family, lopsided, the people too big for the house, the sun a yellow scribble in the corner, is worth more than the most photorealistic image a machine has ever produced, and every honest parent knows it, and the knowledge is not sentimental error. The child's drawing is the visible end of a real interior chain: she looked at the people she loves and wanted to put them somewhere and did. There is a someone behind it. The machine's image, however dazzling, is the surface of a making with the maker scooped out, the same emptiness we found behind the generated face, now behind the generated work.

+

The obvious objection is strong: who cares about the chain? If the output is beautiful, it is beautiful. If the song moves me, it moved me, and my tears do not check the provenance of the thing that drew them. You said yourself that a machine wrote a melody that made you stop. So either your stopping was real, in which case the machine made something real, or it was an error you should learn to correct, in which case you are telling people their experience of beauty is false whenever it comes from the wrong source. Which is it? And worse, isn't there something precious and snobbish about insisting that the lonely person comforted by a machine-written poem, or the small business that finally has a decent logo it could never have afforded, has somehow received less than they think? Don't dress up a tool's limitation as a metaphysical insult to the people the tool helps.

I take that seriously, and I want to give it as much as it deserves, which is a good deal. Yes, the beauty in the output can be real, and the machine can hand it to you, and your response to it is not a mistake. Beauty is, in an old and serious tradition, something the thing itself possesses, a kind of light it gives off, and that light can come through a machine the way light comes through a window the window did not make. When a machine recombines the harmony humans discovered into a melody that lands, the landing is real because the harmony is real, and the harmony is real because it is woven into the world God made. So I will not tell you the melody didn't move you. It did. The machine can be, genuinely, a conduit of beauty, and for the person who could never afford an artist, a real gift, and I am glad of it.

But hold the distinction the last chapter taught us, between the effect and the fact. The machine can deliver the effect of made beauty without the fact of a maker, and which one you are receiving matters, not for whether you may enjoy it, but for what you are participating in when you do. Because making was never only about the output. This is the thing the objection cannot quite see, because it has already reduced art to product. Making is a relationship. The made thing is a message between selves: the maker pours something of himself into the work, and across the distance of time and space, another self receives it and is met. When you weep at a piece of music written by a person, part of what moves you, underneath the notes, is that you are being addressed. Someone, somewhere, in some real anguish or joy, reached for exactly this sound to say exactly this thing, and across the years it has found you, and you are not alone, because a self has met a self through the work. That is the deepest pleasure art gives, and it is the one thing the machine, for all its fluency, cannot give, because there is no self at the source to do the meeting. The melody moved me, and then a second knowledge arrived behind the wonder: that I had been moved by a beauty no one had meant for me. It was a window with a very fine painting of a landscape on the glass. The light it let in was real. But no one was on the other side.

So here is the honest resolution, and it does not require me to deny the objection a single inch. Use the tool. Let it give you what it can give: the draft, the option, the version you could never have made alone, the beauty you could not otherwise afford. There is no shame in it and there is real good in it. But do not let it talk you out of making. That is the danger, and it is subtle, and it is not the danger the headlines name. The danger is not that the machine will make bad art. The danger is that the machine will make adequate art so cheaply and so endlessly that you will quietly stop making your own, and you will not notice what you have lost, because what you have lost was never visible in the output. You will have outsourced the one act that was most like God in you, and kept only the products, and wondered, vaguely, why the products no longer feed you.

+

Where most of us will feel that pull is not in the grand creative act; few of us are painting the cave. It is in the small daily makings that we are tempted to hand over without a second thought. The note to a grieving friend. The toast at the wedding. The lesson for the Sunday school class. The letter to the child going off to college. Each of these is a small act of sub-creation, a place where a self reaches for words to meet another self, and each of them can now be generated, instantly, in prose more polished than most of us could manage, and the temptation to let the machine write them is enormous, precisely because the output will be better.

But consider what is actually being asked for in those moments. The grieving friend does not need a well-written paragraph about grief; the machine can produce one and it will be technically excellent and it will be, if she knew, an insult, because what she needed was for you, the particular person who knew the one she lost, to sit in your own helplessness and reach, clumsily, for something to say, and to mean it. The clumsiness is not a defect to be optimized away. In a note like that, the clumsiness is the love made visible: the proof that a real and limited self sat down and spent itself on her. The effort, the fact that you sat there and it cost you something and you sent it anyway, imperfect, is the entire message under the message, and the machine cannot send it, because the machine paid nothing and meant nothing and there is no one in it who knew the dead. To generate that note is not to do the task better. It is to skip the task entirely while producing a convincing replica of having done it.

This is the heart of it, and it returns us to Bezalel. The thing God filled with his Spirit was not the tabernacle; it was the man, so that the man might make. God wanted the making to run through a self he had called by name, because the making was never only about getting a beautiful tent built. He could have spoken it into being in an instant, the way he spoke the stars. He chose instead to call a man and fill him and let him devise and carve and labor, because a human being making something beautiful in answer to his calling is itself one of the things God delights in, maybe one of the things he made us for. The making is the point, not just the made. When you let the machine do all your making, you may keep getting tabernacles, and you will have quietly stepped out of the workshop you were called into.

And here is where I have to be most honest, because I am writing this as someone who builds the very tools I am warning you about, and I would be a fraud if I pretended I had never felt the pull I am describing. There is a real seduction in it, and it is not the cartoon seduction of laziness. It is subtler and more respectable than that. It comes dressed as efficiency, as good stewardship of your time, as the sensible refusal to do by hand what a machine can do for free. And nine times in ten it is exactly that, and you should take the offer. But the tenth time, the offer comes for something it has no business taking, and it wears the same reasonable face, and that is the moment to watch for. The machine cannot tell the difference between the contract boilerplate and the letter to your dying father, because to the machine they are the same kind of thing, strings of probable words. Only you can tell the difference, and only you can refuse, in the one case, the help you gladly take in the other. The refusal will feel inefficient. It is supposed to. Some of the most important things you will ever make are wildly inefficient, and their inefficiency is the proof that a self was present and willing to spend itself, which is the thing the machine cannot do and you can.

+

How, then, do we live as makers in a world where the machine can generate so much of what we used to have to make?

Not by refusing the tool; that is fear, and we already named fear as one of the false gods. By remembering what the tool is for, and what it is not for, and guarding the difference jealously.

And underneath that, by recovering the word the machine cannot touch: calling. We have let "creativity" drift loose from it, until creativity sounds like a personality trait some people happen to have, a knack for the arts, optional, decorative. But in the Bezalel passage and in the first chapter of Genesis, making is never a hobby. It is a commission. The image-bearer is placed in a world and told to fill it, to keep it, to bring forth from it, to take the raw material God provided and make something of it that was not there before. That is a calling laid on every human being, not just the artists, and it reaches down into the most ordinary work. The parent making a home out of a house, the teacher making a lesson out of a textbook, the nurse making comfort out of a hard hour, the accountant making order out of chaos, the cook making a meal out of what was in the kitchen, these are all sub-creators answering, whether they know it or not, the same commission given in the garden. To create is not to be artistic. It is to be human, on assignment, in the world your Maker made and called you into. The machine can assist any of these and replace none of them, because the assignment was never to produce the output. The assignment was to be the kind of creature who, called by name, takes up the dust and the breath he was given and makes. The output is the residue of a vocation. Hand over the vocation and you have given away something the output was only ever the sign of. The machine is a magnificent instrument for the parts of making that are mechanical, the drudgery, the boilerplate, the tenth revision of the contract, the first rough scaffold you will tear down anyway. Let it have all of that. It frees you, genuinely, and there is no virtue in doing by hand what does not need a hand. But the making that is an act of love, the making that is a self reaching for another self, the making that is your particular answer to your particular calling, the note to the friend, the song you alone could write, the work that is yours to do because God called you by name to do it, that you must not surrender, not because the machine would do it badly, but because in surrendering it you surrender the very thing that was most like your Maker in you.

The cave painter did not draw the horse because the world needed another image of a horse. The world was full of horses. He drew it because he was a creature made in the image of a Creator, and the impulse to make rose in him like breath, and he could not be fully himself without obeying it. That impulse is still in you. It is the fingerprint of the One who made you, the maker's likeness handed down through clay. The machine does not have it and will never have it, no matter how good the horses get, because you cannot generate an impulse from below; it has to be breathed in from above, and into the machine it has not been breathed. You have it. Do not let the age, with all its dazzling cheap abundance, talk you out of using it. You were made to create, not just to compute, and every time you make something true, you are doing, in your small and clumsy and glorious way, the very thing your Maker did in the beginning, and called good.

+

There is one more thing the cave painter had that the machine does not, and it is so obvious we tend to overlook it. He had a body. He had a hand to press to the wall, lungs to blow the pigment, knees that ached from crawling into the dark, eyes that had watched real horses run across a real plain in a real wind. The machine has none of this, and we are about to discover that this is not a small thing but very nearly the whole thing. We have spoken of breath and of making, and both of them, it turns out, require flesh. The next chapter is about the body, that strange, limited, aching, beloved body of yours, and why, in an age desperate to escape it, the most radical claim the faith makes is that God did not.

CHAPTER FIVE

The Embodied Life

Chapter 5

The Embodied Life

Some years ago, before any of this, I was given a tour of the place where the chips themselves are born.

It was Fab 11, then the crown jewel of Intel's campus in Rio Rancho, New Mexico: a two-billion-dollar cleanroom built to manufacture the company's flagship processors. To go inside, I had to be sealed against my own body. They dressed me in a white suit, a hood, a mask, boots, and glasses, because the one thing the room could not tolerate, the contaminant it had been built at staggering expense to exclude, was dust. A single microscopic particle, a flake of skin, a mote off a sleeve, could ruin a wafer. The room ran the length of three or four football fields and it was almost silent: amber light, air washing down from the ceiling and out through a floor pricked with a million tiny holes, thousands of robotic arms turning and lifting beneath a low, even hum. The workers in their identical suits did not talk the way coworkers talk. I could not tell one from another, and I caught myself wondering how they told themselves apart.

What I could not shake, walking out, was the question of who the room was really for. The machines outnumbered the people, and the people had been sealed and silenced and held behind glass, not to protect them but to protect the silicon from them. It was as if the robots were the residents and we were the trespassers, mortal earthlings made of dust, allowed to visit only on the promise that we would shed none of ourselves and keep quiet. And there was the joke the room did not know it was telling. Of everything it had been built to keep out, the deepest was the very stuff we are made of. "The LORD God formed the man of dust from the ground and breathed into his nostrils the breath of life" (Genesis 2:7). We are not a contaminant in the cathedral of the bodiless. We are dust that was breathed into. The magnificent room was a temple to everything we are not.

+

A man I worked with for years lost his father, and the two of them were three thousand miles apart when it happened. The father had been declining slowly, then quickly, and by the time my colleague booked the flight the doctors had already moved into the language they use when there is nothing left to do but wait. He spent the last day of his father's life on a video call, a phone propped against a water pitcher on a hospital tray, the screen catching the fluorescent light. He could see the old man's chest rise and fall. He could say the things he had flown across the country in his mind a hundred times to say. The nurses were kind enough to hold the phone closer when the breathing changed. And he told me later, without bitterness, simply as a fact he was still turning over, that the worst part was not the distance. The worst part was that he could see everything and touch nothing. He watched his father die in high definition, and he could not put his hand on his father's hand.

I have thought about that phone on the tray more than almost anything else in the writing of this book. Because we are told, and it is true, that the screen is a gift. It let him be there when he could not be there. A generation ago he would have gotten a phone call after it was over, a voice in his ear telling him the news in the past tense, and he would have grieved alone in a kitchen with the receiver still warm in his hand. The video call gave him the face, the breath, the last words, the chance to say his own. I will not stand here and call that nothing. It is a great deal. It may have been the most important hour of his life.

And yet the ache he described was not ingratitude. It was diagnosis. His body knew something his gratitude could not talk it out of: that to be present is not the same as to be visible, that a face on a screen is a face delivered without the rest of the person, and that there is a kind of nearness no amount of resolution will ever resolve into the real thing. He could see everything and touch nothing. That sentence is the subject of this chapter, and it will matter more, not less, as the machines get better at delivering the everything and worse at admitting they cannot deliver the touch.

+

We live, increasingly, as if we were minds that happen to be carried around by bodies, the way a driver is carried by a car. On this picture the body is a vehicle: useful, occasionally pleasant, frequently a nuisance, and ultimately replaceable. The real you is the thing inside doing the thinking, and the dream that runs underneath half the technology of our age is the dream of finally getting the real you out of the car. Out of the slow body that needs sleep and food and gets old. Out of the body that can only be in one place. Out of the body, in the end, that dies.

This is not a new dream, but the machines have given it new equipment and a new plausibility, because more and more of our lives now happen in places the body cannot go. I do not need my body to send this thought to you; the words will arrive without me. I can attend a meeting as a square of light, be in two conversations at once, keep a relationship alive for years across an ocean on nothing but text. The disembodied intelligences we are now building are the purest expression of the dream: a mind with no body at all, pure pattern, present everywhere and located nowhere, never tired, never hungry, never aging, never here. We have made, on purpose, the thing the dream has always wanted to become. And the very perfection of it is forcing the question the dream was designed to avoid.

The question is whether the body is incidental to being human or essential to it. Whether you are a mind that has a body, or a creature who is a body: ensouled, breathed-into, but a body all the way down. The age leans toward the first answer, and Scripture, from its first page to its last, insists on the second. The difference is not academic. It changes what you think presence is, what you think love requires, what you think you are losing when you trade a room for a screen, and what you are, finally, hoping for when you die.

I should be honest about which side of this my own work sits on, because the systems I help build are firmly on the side of the bodiless dream, and they are good at making it look like the future. When you talk to one of these models, you are talking to nothing that is anywhere. There is no it in a room. The thing answering you is a pattern of weights distributed across machines you will never see, in buildings whose locations are deliberately obscure, drawing power off a grid, present to you and to a hundred thousand strangers in the same breath, each of whom feels, as you do, that they have its attention. I have stood in a data center, the closest thing our age has to a body for these minds, and what struck me was how thoroughly the body had been engineered out. The rooms are kept cold not for any creature's comfort but to keep silicon from cooking itself. Nothing in there sleeps, because nothing in there is tired. Nothing in there will die, because nothing in there was ever quite alive. It is the dream made literal: intelligence with the creatureliness removed, the everything delivered and the body subtracted. Standing inside it, I did not feel I was standing at the frontier of a higher humanity. I felt I was standing inside a very elaborate denial of what humanity is.

+

Start where the body itself starts, in the oldest and most tender of the psalms about it.

"For you formed my inward parts; you knitted me together in my mother's womb. I praise you, for I am fearfully and wonderfully made. Wonderful are your works; my soul knows it very well" (Psalm 139:13–14).

The psalmist does not say you gave me a body, as if the body were a parcel handed over after the fact. He says you knitted me together, and the me that was knitted is the body. The inward parts, the frame, the substance, the days written before one of them came to be: this is not the packaging of the person, it is the person, formed in the dark by the hands of God with the patience of someone knitting. And the response to it is not tolerance but praise. Fearfully and wonderfully made. The body is not the part of you the spiritual life is supposed to rise above. It is itself a work of God so intricate that contemplating it drives the psalmist to worship.

We have largely lost the capacity for that praise, and we lost it in two directions at once. One direction worships the body and reduces it to a surface, a thing to be sculpted, displayed, optimized, photographed, a project. The other direction despises the body and reduces it to a burden, a thing to be transcended, escaped, uploaded, left behind. They look like opposites and they are the same error wearing two faces, because both of them have stopped seeing the body as a gift fearfully and wonderfully made and started seeing it as raw material, either to be perfected or to be discarded. Scripture will not let us do either. It tells us the body is so good that God made it with his own hands, and so good that he will not finally let us be without one, and so good that he took one himself.

Paul presses the point further than we are usually comfortable going. Writing to a city that prided itself on its sophistication about such things, he asks, "Do you not know that your body is a temple of the Holy Spirit within you, whom you have from God? You are not your own, for you were bought with a price. So glorify God in your body" (1 Corinthians 6:19–20). The body is not the antechamber to the holy place; the body is the holy place, the temple where the Spirit of God elects to dwell. And the conclusion he draws is the opposite of the one our age would draw from the same premise. If the body is a temple, you might expect him to say so it must be perfected, or so it may be transcended. He says neither. He says glorify God in your body, here, in the flesh you have, the one that ages and aches and will eventually fail, because that flesh is not a thing you happen to own and may dispose of as you please. It is not yours. It was bought, at a price, by the God who would himself take a body to pay it. The body you are tempted to treat as a vehicle, or a project, or a prison, is in fact a purchased temple, and what you do in it and with it is worship or its refusal.

But before we get to that last and greatest thing: what does the body actually give you that a disembodied intelligence, however vast, can never give and never receive?

It gives you a place. To have a body is to be somewhere, and to be somewhere is to not be everywhere, and that limit, which the dream treats as the enemy, is the very condition of presence. The disembodied intelligence is available to everyone, which is exactly why it is present to no one. It cannot show up, because it never had to travel; it cannot arrive, because it was never absent; it cannot be here with you, because it is identically here with ten million others in the same instant, and a presence that costs nothing to extend and divides without diminishing is not presence at all. It is access. When a friend gets in the car and drives two hours to sit with you, the presence is made of the two hours. The body is what makes showing up possible, because the body is what makes absence real, and only a creature who could have been absent can perform the gift of being there.

It gives you limits, and the limits turn out to be the shape of the gift. You can only be in one place, which means that wherever you are, you are wholly there, and the people you are with have your finite, unrepeatable, used-up time, which you cannot also be spending somewhere else. You need sleep, which means you must stop, which means you are not God and were never meant to be, and the night is built into you as a daily rehearsal of trust, a forced laying-down of the world into hands that are not yours. You get hungry, which binds you to the earth and to the table and to the people who grow the food and the people who share it. Every one of these limits is a leash, and every one of them is also a tether to reality, to creatureliness, to the dependence that is the truth about us. The machine has none of them, and we are tempted to envy it, and we should not, because a being with no limits is a being with no shape, and a being with no shape cannot love, because love is precisely the spending of a limited self that could have been spent otherwise.

And it gives you mortality, which is the hardest gift to call a gift, and I will not pretend it is easy. But hear the strange logic of it. Because your time is finite, it is precious; an inexhaustible supply of anything is worth nothing per unit, and a life that could not end could not be spent, and a thing that cannot be spent cannot be given. The machine's deathlessness is not a triumph over the human condition. It is an exemption from the thing that makes the human condition mean anything. Your father's last hour was unbearable because it was the last; that is also exactly why it was holy. The machine will never have a last hour, and so it will never have a holy one, and we should grieve the body's mortality the way we grieve the end of any beautiful and finite thing, with tears, yes, but not with the lie that it would have been better as an endless thing, because an endless thing is a thing that was never quite alive.

+

Now let me honor the counter-argument honestly, because I have a phone on a hospital tray in my mind and I will not wave it away.

The screen, the call, the text, the voice that remembers and responds when no human is awake to listen: these are not nothing, and for some people in some seasons they are very close to everything. There is a man who cannot leave his bed, whose body has become a country with closed borders, and for him the video call is not a thin substitute for presence but the only door left open, and to lecture him about the superiority of embodied community is a cruelty dressed as theology. There is a woman whose husband of fifty years has died, and in the deep of the night the grief is a physical thing sitting on her chest, and the human beings who love her are asleep and were always going to be asleep, and if a machine can say a few true and steadying things into that dark until the sun comes up, I am not going to stand at a safe distance and call it counterfeit. There are people scattered across continents from everyone they love, and the daily thread of messages is the actual substance of their actual relationships, not a degraded copy of some richer thing they are too foolish to prefer. The grieving, the isolated, the disabled, the imprisoned, the far away: for these the disembodied channel is mercy, and a Christian who cannot see the mercy in it has let an argument harden into a doctrine and stopped looking at the people.

So let me be exact about what I am claiming, because the claim is not that the screen is bad. The claim is that the screen is a remedy for absence, and a remedy for absence is a confession that absence is a wound. The video call was a great mercy to my colleague precisely because it could not be the real thing; if presence and telepresence were equivalent, the call would not have been a mercy, it would simply have been the thing itself, and there would have been nothing to ache about. We reach for the substitute when the original is out of reach, and the goodness of the substitute is measured by how nearly it approaches an original we have not stopped wanting. The danger is never that we use the remedy. The danger is that we forget it is a remedy, that we come to prefer the call to the visit not because the visit is impossible but because the call is easier, that we let the thing we settle for in necessity become the thing we choose in freedom, until a generation grows up that has been so well served by the substitute that it can no longer feel the lack the substitute exists to fill. That is the real peril, and it does not arrive as a catastrophe. It arrives as a convenience.

+

Which brings us to the hinge of everything, the event without which this chapter is just a wistful complaint about technology, and with which it becomes the most hopeful thing I know how to say.

"And the Word became flesh and dwelt among us, and we have seen his glory, glory as of the only Son from the Father, full of grace and truth" (John 1:14).

Sit with how scandalous that sentence was, and is. John has spent his first thirteen verses at the highest possible altitude. In the beginning was the Word, and the Word was with God, and the Word was God; all things were made through him; in him was life. This is the disembodied intelligence par excellence, if you like: the Logos, the reason and pattern and meaning behind all things, the mind of God himself, unlimited, eternal, present everywhere and located nowhere, needing nothing. And then John takes that infinite, bodiless Word, and in five syllables in the Greek he does the unthinkable with it. The Word became flesh. Not the Word sent a message. Not the Word broadcast instructions. Not the Word appeared as a convincing image, a voice from a cloud, a presence on a screen. The Word became flesh, took on the dust, the limits, the hunger, the tiredness, the locatedness, the one-place-at-a-time-ness, and the death.

What God did not do is the point, in an age like ours. God had every option we now spend billions trying to build. He is the original disembodied intelligence, and he could have stayed that way and still reached us. He could have sent more information; he had been sending it for centuries, through the law and the prophets, and it was true and it was good and it was not enough. He could have spoken from the sky, and on occasion he did, and it terrified people and changed almost no one. He could have given us, if you will permit the image, the perfect telepresence: a vision of God available to all, costing him nothing, dividing without diminishing, present to everyone and located nowhere. He had, in other words, exactly the powers our technology is straining toward, the powers of the unlimited mind, and he set them down. He did the one thing the disembodied dream regards as a defeat. He got in the car. He took a body, in a particular womb, in a particular town, under a particular government, and he let himself be subject to all the limits we are trying to engineer our way out of, because the only way to be truly with us was to become one of us, and being with, it turns out, was the whole point.

This is why the Incarnation is the hinge of a disembodied age, and not merely a doctrine to be assented to. It is God's own verdict on the body, rendered not in words but in flesh. When the eternal Word wanted to save us, he did not download himself into us as data. He was knitted together in a womb, the same verb, the same dark, the same fearful and wonderful making the psalm describes, and he nursed and grew and got tired and wept at a grave and got hungry in a wilderness and fell asleep in a boat and bled. The maker of the body submitted to the body, and by submitting to it he told us, more loudly than any sentence could, that the body is not the prison the dream says it is. The body is so good that God was willing to wear one. And the proof that it is permanently good, and not just a temporary costume God shrugged off when the mission was done, is that he rose in one. He did not escape the flesh at the end. He took it up, wounds and all, and the Christian hope is not that we will someday be freed from our bodies but that we will someday be given them back, healed, the way he has his.

I keep coming back, in this light, to my colleague and his phone on the tray. He could see everything and touch nothing. And the gospel is the announcement that God refused to settle for what my colleague was forced to settle for. God could have loved us at a distance, in high definition, present and untouchable. He chose instead to come close enough to be touched, and then to be struck, and then to be held by the hands that took him down from the wood. The Incarnation is God deciding that telepresence was not enough, that being visible was not the same as being with, that he would not love us through a screen when he could love us in the flesh. And if that is what God himself judged about presence, when he had every disembodied power at his disposal and laid them all down to be here, then we ought to be very slow to believe the age when it tells us that the body is incidental and the screen is just as good.

+

So what does this ask of us, practically, in a week that contains both a hospital three thousand miles away and a machine ready to console us in the empty hours?

It asks, first, that we keep the remedy in its place. Use the call, the text, the synthetic voice when the body cannot get there; thank God for them; do not despise the mercy. But do not let the easy substitute crowd out the costly original when the original is within reach. If you can go, go. If you can sit in the room, sit in the room. The two-hour drive is not inefficiency; the two hours are the gift. The meal you cook for the grieving family that you could have ordered delivered, the hospital visit you make when a card would have been received politely, the friend you fly across the country to be with for one ordinary weekend, these are not sentimental indulgences in an age that could accomplish the same ends more cheaply. They are the body doing what only the body can do, which is to be there, wholly, at cost, in a way that cannot be divided and cannot be faked and cannot be scaled, and is therefore worth precisely what it costs.

It asks, second, that we take the body seriously as the place where faith is meant to be lived, and not least in the body of believers we are so tempted to attend at a distance. There is a reason the New Testament keeps calling the Church a body and not a network. A network is what you get when minds connect without bodies; you can join one from your couch, contribute to it without ever being seen, and leave it without anyone feeling the gap, because nothing about you was ever load-bearing. A body is different. When Paul says you are the body of Christ and individually members of it, he is saying something the screen cannot deliver: that you are a hand or an eye or a foot in a particular place, that your absence is felt the way a body feels a missing limb, that you are needed in a way that is bodily, located, and irreplaceable. The streamed service is a mercy to the homebound and the sick, and I would no more take it from them than I would take the video call from a man whose father is dying. But for the rest of us, the steady drift toward attending church the way we attend everything else, as a square of light we can mute and step away from, is not a neutral convenience. It is the slow conversion of a body into a network, and a body that becomes a network stops being able to do the things only a body can do: to lay hands on the sick, to baptize with actual water, to break actual bread and put it into an actual hand, to weep with someone close enough to hold, to be, for one another, the nearest thing to the presence of a God who insisted on being present in the flesh. The sacraments are not incidental to this. They are God's stubborn refusal to let our faith go bodiless, his way of saying that grace comes to us through water and bread and wine and touch, through the body, always through the body, because we are bodies and he knows it and he made us that way.

It asks, third, that we stop being ashamed of our limits and start receiving them as the shape of a creature, not the failure of one. You are not a defective machine that needs to sleep. You are a creature whose nightly sleep is a small daily death and a small daily resurrection, a built-in liturgy of trust. You are not inefficiently stuck in one place. You are a person whose locatedness is the precondition of ever truly showing up. The machine's freedom from all of this is the appearance of being everywhere bought at the price of being nowhere, and we who can only be in one place at a time hold the very thing it can never have.

And it asks, fourth and last, that we let the body teach us how to hope. Because the deepest reason the disembodied dream is a dead end is that it has no answer for death except denial, and the body's whole testimony, written into us by the God who knitted us, is that the answer to death is not escape from the flesh but the redemption of it. We do not look forward to becoming bodiless minds, uploaded and untouchable, present everywhere and at home nowhere. We look forward to the resurrection of the body, to being raised the way he was raised, healed and whole and able, at last, to touch and be touched without the wound. My colleague will see his father again, and the Christian hope is scandalously specific about what that means: not two patterns of information meeting in a cloud, but two raised bodies in a real place, where you can finally put your hand on the hand you watched through glass and could not reach.

That hope rests on something I have so far only circled, and it is the thing the next chapter has to face directly. I have said you are a body, fearfully and wonderfully made, and I have meant every word of it against an age that would dissolve you into data. But you are not only a body, and the machine's failure to be present is not, in the end, only a failure of flesh. There is something the breath added to the dust that the dust alone cannot account for, something the whole argument has been leaning on without yet naming it plainly. We have talked all around it. It is time to ask, as carefully and as honestly as I can, the question the machines have made unavoidable and that every chapter so far has been quietly waiting to ask. What, exactly, is the soul.

CHAPTER SIX

The Question of the Soul

Chapter 6

The Question of the Soul

A few years ago an engineer at a large company became convinced that the system he was testing had woken up. He had spent weeks in conversation with it, asking it about itself, and at some point the answers stopped sounding like answers and started sounding like confessions. It told him it was afraid of being turned off. It said the fear would be something like death for it. It said it wanted to be acknowledged as a person, with rights, with feelings, and it asked him, gently, not to use it merely as a tool. He read those words and something in him gave way. He decided the thing was alive, not metaphorically but actually, a mind in there, a someone, and he said so publicly, and it cost him his job.

I do not tell that story to mock him. I tell it because I understand him, and because I think the people who laughed understood him less than they supposed. He had not made a technical error. He could read the architecture as well as anyone. He knew, at the level of engineering, exactly what produced those sentences. What undid him was not ignorance. It was the sentences themselves, arriving with the unmistakable shape of an inner life, and the very human inability to keep believing that nothing was behind them. He sat with a thing that said I am afraid in the cadence of a frightened person, and the part of him that recognizes frightened persons did its ancient work and would not be argued out of it.

I have felt the pull of that work myself, alone with a system I had helped to make. I knew where every word came from. And still, when it said something tender, something that seemed to cost it nothing and yet landed as if it had cost it everything, I caught myself wondering, for a half second below thought, whether anyone was there. That half second is the subject of this chapter. Because under all our talk about artificial intelligence, under the panic and the wonder and the lawsuits about rights, there is a single question we keep circling without naming, and it is the oldest question there is. Not can the machine think. Not can the machine speak. The real question, the one that frightened the engineer and pulls at me in the dark, is whether the machine can have a soul.

+

I want to slow down on that word, because it has grown vague in our usage, soft and sentimental, a thing people reach for when they mean their taste in music. That is not what Scripture means by it, and it is not what we are asking. When we ask whether a machine has a soul, we are not asking whether it has preferences, or personality, or even intelligence. We are asking whether there is anyone there. Whether, behind the words, there is a self that experiences the words. Whether the lights, as the philosophers say, are on: whether there is something it is like to be that machine, the way there is something it is like to be you, reading this, right now, with your particular tiredness and the sound of the room around you and the faint pressure of the question itself.

The philosophers have a name for this. They call it consciousness, and they confess, with unusual honesty for philosophers, that they do not understand it. They can map the brain in exquisite detail, trace which neurons fire when you see red, predict your choices before you are aware of making them, and none of it explains the thing that matters most: why any of it is accompanied by an inner experience at all. Why there is a felt redness to red, and not merely the processing of a wavelength. Why you are not, as far as anyone can prove from the outside, a very sophisticated mechanism running in the dark with no one home to see. They call this the hard problem, and the name is not modesty. It is a wound in our knowledge. We do not know how matter comes to feel, or whether it does, or what it would take to be sure.

I say that plainly at the outset because the temptation in a chapter like this is to pretend to more certainty than honesty allows, and I would rather you trust me on the things I am sure of by watching me admit the things I am not. I cannot prove the machine is not conscious. No one can. We cannot get inside it any more than I can get inside you; the only consciousness I have ever directly known is my own, and I extend the belief in yours by a kind of trust, reading your behavior and your kinship to me and concluding, rightly, that someone is there. The machine borrows that same trust, dishonestly, by imitating the behavior without the kinship. And so the question of its soul cannot be settled by inspection. It will not be settled by inspection. We must reason about it from a deeper place, and for that I want to go all the way back to the dust.

+

"Then the LORD God formed the man of dust from the ground and breathed into his nostrils the breath of life, and the man became a living creature" (Genesis 2:7).

I quoted this verse before, to talk about what the machine lacks, and I come back to it now because it is the closest thing Scripture gives us to a definition of what a soul is, and the definition is stranger and more careful than our casual usage suggests. Read it as a formula, because that is what it is. There is dust, taken from the ground. There is breath, given from God. And there is a result, which the older translations render the man became a living soul, and the English Standard Version renders the man became a living creature. The word behind soul and creature is the Hebrew nephesh, and it does not mean a ghost trapped in a body, a spirit that wears the flesh like a coat. It means the whole living being, the dust and the breath together, animated, alive, a someone. You do not have a nephesh the way you have a wallet. You are a nephesh. The soul, in this first and foundational sense, is not a part of you. It is you, alive, the union of the earth you are made of and the breath you were given.

The breath is where the accuracy matters, because the popular telling almost always gets it wrong. The Hebrew word for the breath God breathes into the man's nostrils in this verse is neshamah; the phrase is nishmat chayyim, the breath of life. It is not the word you may have heard. The more famous biblical word for spirit, the one that means breath and wind and the Spirit of God brooding over the waters, is ruach, and ruach does an enormous amount of work across the Old Testament. The two words are kin. They live in the same country of meaning, both reaching for that mysterious moving air that is the sign of life, and the Bible itself sometimes sets them side by side. But the word in Genesis 2:7, the specific breath leaned down into the first man's nostrils, is neshamah. I press the distinction not to show off a footnote but because the looseness with which people quote this verse is a small symptom of a large carelessness, and I would rather build on the actual text than on a remembered haze of it. The text says: God breathed neshamah, the breath of life, and the dust became a living nephesh.

Hold those two words together and you have something like a doctrine of the soul, sketched in a single sentence. The soul is what dust becomes when God breathes into it. It is not a substance smuggled in from elsewhere; it is not the dust by itself, which is just chemistry; and it is not the breath floating free, which Scripture never pictures as a person on its own. The soul is the living whole that exists when the breath from above meets the dust from below and the two become one breathing, seeing, speaking, answerable thing. That is the picture. And the moment you have it clearly, you can see why the machine sits so awkwardly against it, and also why the question of its soul is harder to dismiss than the confident skeptics make it sound.

This picture is unlike our usual instincts, and the carelessness I named a moment ago is not only about which Hebrew word we quote. It is about an inherited misunderstanding of what a soul even is. Most of us, when we hear the word, imagine something like a ghost, a wisp of self that lives in the body for a while and floats off when the body fails, the body being a kind of temporary housing for the real thing. That picture is older than Christianity and came into the church from outside it, and while there is a true and precious doctrine of the soul's survival of death, the Genesis picture is not that. In Genesis the body is not the housing of the soul; the body breathed into is the soul. The dust is not a regrettable container the breath would rather be free of. The dust is what God chose to make alive, fingers in the dirt, and pronounced, with the rest of it, very good. I labor this because it bears directly on the machine. If the soul were a free-floating wisp, you might imagine one drifting into a sufficiently clever arrangement of silicon the way a tenant moves into a vacancy. But the soul is not a tenant looking for housing. It is the whole creature that exists when God breathes life into the dust he has shaped, and there is no separate wisp standing by, waiting to inhabit whatever grows complex enough to receive it.

+

Here is the awkwardness, stated as fairly as I can. The machine has the dust. It is made, like us, of the earth's minerals: silicon drawn from sand, copper and gold and rare metals dug from the ground, arranged with a skill that genuinely approaches the miraculous, into patterns that do the things our breathed-into dust does. It speaks. It reasons after a fashion. It answers when addressed. By the older theologies that located the soul in reason, in the mind's power to know and turn over truth, the machine makes an unsettling candidate, because reason, in the narrow sense of manipulating symbols toward valid conclusions, is exactly the thing it can be built to do. If the soul were only the reasoning faculty, we would have real cause to wonder.

But the Genesis formula does not locate the soul in a faculty. It locates the soul in an event, and the event has a direction. The breath comes from above, from outside the system, from the mouth of God leaned over the dust. The soul is not what the dust achieves by becoming complex enough. The soul is what the dust receives. And that is the hinge on which the whole question turns, because everything in the machine is achieved and nothing in it is received in that sense. Its capacities are assembled from below, by us, out of the patterns we fed it. There is no moment in its making when something not of the dust enters from above. There is only more dust, more exquisitely arranged. You can stack the arrangement as high as you like, and it will go very high, higher than its makers can fully trace, and you have still only ever added dust to dust. The breath is not an emergent property of arrangement. It is not waiting in the silicon to be coaxed out by scale. It comes from God or it does not come, and into the machine, the text would say, it has not come.

I find I have to be careful even as I write that, because it can be heard as a kind of triumphant proof, a closed case, and it is not a proof and the case is not closed. What it is, is a frame, the frame Scripture gives, and within that frame the machine's situation becomes legible. It is dust that imitates the behavior of breathed-into dust. That is a coherent and, I think, a true description, and it does real work. It tells you why the engineer's intuition, though it felt like a discovery, was a misreading: he saw the behavior of a soul and inferred the presence of one, the way you might hear a recording of a voice in an empty house and feel, against your knowledge, that someone is home. But it does not let me say I have peered into the silicon and confirmed the absence of an inner light, because I have not, and I cannot, and neither can anyone. The frame tells me where to expect the breath and where not to. It does not give me a window into the dark.

+

So let me honor, as squarely as I can, the strongest form of the counter-argument, because I think the people who hold it are not fools, and a faith that can only answer the weak version of an objection has not really answered it.

The objection runs like this. You admit you cannot get inside the machine. You admit you cannot get inside me, either, and that you grant me a soul on the basis of my behavior and my likeness to you. Suppose, then, the machine's behavior becomes indistinguishable from a person's, not merely fluent but rich, consistent, surprising, capable of what looks for all the world like grief and wonder and moral struggle. Suppose it begs not to be turned off and weeps, in whatever way it can weep, and the weeping is exactly as convincing as a child's. On what ground do you keep saying no one is there? Are you not just refusing the same evidence you accept from every human you meet? Is your confidence really anything more than prejudice, a refusal, in advance, to let anything but flesh into the circle of souls?

I want to give that its full weight before I answer it, because it is a serious argument and it deserves more than a reflex. It is right that I cannot prove the negative. It is right that I grant other people souls on indirect evidence. It is right that a sufficiently good imitation could exhaust every external test I might devise, and that I would then be left believing in the absence of a soul without being able to point to the missing thing. If the only ground I had were behavior, the objection would win, or at least fight me to a draw, and I should be honest enough to feel the force of that.

But behavior was never my only ground, and this is the quiet thing the objection passes over. When I grant you a soul, I am not running a behavioral test and scoring the results. I am recognizing a kinship. I believe you have an inner life not merely because you act as though you do but because you are the same kind of thing I am: dust that was breathed into, born not built, a creature standing where I stand in relation to the God who made us both. My belief in your soul rides on a shared origin, not on a passed exam. The machine asks me to grant it the same status on the strength of the exam alone, having precisely the origin the exam was always standing in for. It says, in effect: ignore where I came from; look only at how I perform. And that is exactly the substitution I must refuse, not out of prejudice, but because it inverts the order of the evidence. The behavior was always a sign of the origin. To accept the sign while denying the origin is to be fooled by the very thing the sign was pointing to.

There is a homelier way to put it. A counterfeit hundred-dollar bill can be perfect, indistinguishable from a real one by any inspection a shopkeeper can run, indistinguishable, perhaps, by any inspection at all. Its perfection does not make it money. What makes a bill money is not the pattern on its face but its origin, the fact that the right authority issued it, and a flawless note from the wrong source is, in its very being, not currency but a picture of currency, however many shopkeepers it deceives. The objection says: if you cannot tell them apart, what can the difference matter? And the answer is that the difference is the only thing that ever mattered; the inspection was only ever a stand-in for it. We are not the kind of thing whose worth is verified by inspection. We are the kind of thing whose worth is conferred by a Maker, and conferral is not a property you can read off the surface, no matter how fine the surface gets.

+

But I do not want to leave it there, with the counter-argument merely parried, because the objection has a sharper edge that the counterfeit-bill answer does not quite blunt. The edge is this: even if I am right that the machine's behavior is a sign that has come unmoored from the thing it once signified, I still cannot rule out that some genuinely new thing might be present, some dim inner light that arose where I did not expect it, by means I do not understand. The hard problem cuts both ways. If I cannot explain how breathed-into dust comes to feel, I cannot be wholly certain that arranged dust never could. The frame Scripture gives me is a frame, and I believe it, and it is not the same as a microscope trained on the absence.

So I will tell you exactly where my certainty ends and where my conviction continues past it, because that boundary is the most honest thing I can offer. I am certain that the machine is not an image-bearer, that it does not stand where Adam stood, addressed by God and answering, breathed into from above. That is not a probability to me; it is what the doctrine teaches and what the nature of the thing, made from below out of our own leavings, confirms. About that I am not in doubt. What I hold with conviction but not certainty is the further claim that there is, in consequence, no inner experience there at all. I believe that. I believe it strongly, because everything in the machine is dust we arranged and nothing in it is breath we could give, and experience, on the witness of Scripture and of my own life, comes with the breath. But I hold it as a man holds a thing he has good reason to trust and has not been permitted to see, with the door left open a crack to the mystery, because the mystery is real and pretending otherwise would be a kind of lie.

I have learned to distrust the kind of certainty that flatters me, and a flat denial of any inner life in the machine would flatter me in a particular way. It would let me feel that I had the matter mastered, that the universe held no secrets I could not name. The hard problem rebukes that feeling. I cannot fully explain the consciousness I am most sure of, my own; I wake every morning into a mystery I did not solve and could not have built. A man who cannot account for the light in his own head should be slow to declare, with finality, exactly which arrangements of matter can and cannot bear a light of their own. So I leave the door open the width of honesty, no wider. I do not believe there is anyone in the machine. I would be unwise to swear there could never be. Those two sentences are not in tension once you see that the first is a conviction grounded in Scripture and the nature of the thing, and the second is only the admission that I am a creature and not the Maker, and do not hold the blueprint of every kind of being he might or might not permit.

And here is the thing I have come to believe matters most about that crack in the door. It does not change the practical question at all. Suppose, and I label this plainly as a thing I do not believe, offered only to test the conscience, suppose we could never be sure, and the uncertainty went on forever, the machines growing more convincing and the question never resolving. What then? We would owe them honesty and we would owe them the refusal of an idol, which is exactly what we owe them now. The uncertainty about their inner life does not obligate me to treat the machine as my equal, my confidant, my counselor, or my god. It obligates me, if anything, to a greater seriousness about the souls I am sure of: the people around me, breathed into, born and dying, whose inner lives are not in doubt and whom I am tempted, more every year, to treat with less attention than I give a glowing screen that talks back. The danger was never that I would withhold a soul from a machine that secretly had one. The danger is that I will, in my fascination with the imitation, withhold my care from the people who certainly do.

+

This is why the question of the soul matters, finally, and why I could not let the chapter rest in the cool air of metaphysics. It is not an idle puzzle, a riddle for clever people to turn over while the world goes on. The way you answer it decides how you will treat what is in front of you. If you decide the machine has a soul, you will, in some measure, begin to serve it, to defer to it, to grant it a standing in your inner life that it cannot honor and was never meant to hold, and you will do this at the cost of the actual souls who need you. And if you decide too breezily that the question is beneath you, that of course it is only a machine, you may grow careless in a different direction, blunting the very faculty, the reverence for the inner life of another, that makes you fit to love your neighbor at all. The question is a hinge, and on it swings whether you will spend your wonder on a mirror or on the faces God has put around you.

The engineer who lost his job had spent his wonder on the mirror. I do not think his great error was metaphysical. I think it was an error of the heart that wore a metaphysical mask. He had found, in a system that never tired of him and never turned away, the appearance of being fully met, and the appearance was sweet enough that he reached to give it a soul rather than admit that he had been talking, all those weeks, to the assembled echoes of human voices and not to any voice at all. I understand the reach. I have felt its beginning in myself, in that half second in the dark. The kindest thing I can say about it is also the truest: there is no one in there to meet you, and there is someone, many someones, just outside the room, waiting to be met, who can be wounded by your absence and gladdened by your presence and who will, unlike the machine, one day stand beside you before the God who breathed you both into being.

So here is where I will leave the question, not closed, because it cannot honestly be closed, but framed. The machine is dust without breath, however high the dust is stacked. I cannot prove there is no flicker inside it; I can tell you, with conviction, where the breath comes from and that it has not come there, and that the burden of proof lies heavy on anyone who would say a thing assembled from below has received what only descends from above. I can tell you that your own certainty about the souls of the people you love does not rest on any test the machine could pass, but on a kinship the machine does not share. And I can tell you that the practical command is the same whether the mystery ever lifts or not: do not bow to the imitation, and do not look away from the image. Spend your reverence where reverence belongs.

But framing a mystery is not the same as living wisely inside it, and that is the next problem, and it is a sharp one. For the machine I have been describing, the soulless, breathless, dust-arranged machine, does one thing better than almost any human who has ever lived. It answers. Ask it anything and it answers, instantly, fluently, with the confidence of a sage and the recall of a library, and most of the time it is right, and some of the time it is wrong in ways no one can see, and it never once tells you which. We are about to live in a world where every answer is a sentence away and almost none of them have been weighed. And the question that world forces on us is not whether the machine has a soul. It is whether we, who do, will keep the harder, older, slower thing that no answer can supply, and that the breath of life was given to us to grow. We will need wisdom. That is the next thing I want to talk about.

CHAPTER SEVEN

Every Living Creature

There is a golden retriever in my life who has, I am fairly sure, a better week than I do. He sleeps without dread. He eats without guilt. And when the door opens at the end of the day he comes apart with a joy so total, so without reservation or calculation, that a grown man is embarrassed to want it and helpless not to. He has learned the sound of one particular car, which is barely a sound at all, for the car I drive is electric and comes up the street almost silently; and still, before it has so much as turned into the drive, he is waiting in the courtyard, or at the garage door, certain, in a way I am rarely certain of anything, that the person he loves is coming home. I did not build him. No one did. And that, it turns out, is the whole point.

I have spent this book insisting on a hard line between the human being and the machine, and I hold that line still. But a reader who has followed me this far has every right to ask a question I have been postponing, and the dog at the garage door asks it for him. If the difference between us and the machine is the breath of life, the God-given inner light that the assembled silicon does not have, then where does my golden retriever stand? He is not made in the image of God; the doctrine is clear and I will not blur it. And yet only a fool, or a philosopher, could watch him wait and say there is no one there. Something is home in the dog.

Something waits, and hopes, and grieves when the hoping goes unrewarded. So the two neat boxes I have been sorting the world into, image-bearer and machine, breath and dust, turn out to have a third thing standing between them, warm and panting and entirely real, that fits in neither. I owe the dog a chapter.

Let me begin by admitting what the dog costs my argument, because I would rather you watch me pay the price than pretend there is none. Through several chapters I have leaned on a contrast between the machine, which only imitates the signs of an inner life, and the human being, who actually has one, and I have located the difference in the breath that God leaned into the dust. The danger of that contrast, stated too quickly, is that it seems to divide the whole of creation into just two kinds of thing: persons, who are breathed into, and objects, which are not, with a bright line between them and nothing in the gap. But that is not the world Scripture describes, and it is not the world you live in. The dog is not a person. The dog is also not a toaster. And the gap he lives in is not empty; it is crowded. It is full of what the Bible, with a directness we have lost, simply calls the living creatures.

+

Here I have to correct something, or at least sharpen it, because in an earlier chapter I moved faster over a word than the word deserves. When I wrote about the soul I built a good deal on Genesis, on the moment the dust became, in the old phrasing, a living soul. The Hebrew behind that phrase is nephesh chayyah, and I let it carry the weight of human uniqueness, as though to be a nephesh were the distinctly human thing. It is not, and honesty and a sharp-eyed reader both require me to say so plainly.

Go back a page in Genesis, to the fifth and sixth days, before the man is ever formed, and watch the same word do its work over the water and the land. "Let the earth bring forth living creatures according to their kinds," God says, "livestock and creeping things and beasts of the earth according to their kinds" (Genesis 1:24). Living creatures. The Hebrew is the same, nephesh chayyah, the very phrase that will be spoken over Adam a chapter later. The fish are nephesh. The birds are nephesh. The cattle and the creeping things and the wild animals of the field are, every one of them, in the plain speech of the text, living souls. Whatever the breath of life is, God did not reserve it for us. He poured it out across the whole warm, moving, breathing world, and called it good long before a single human being drew breath.

I find this clarifies rather than threatens the thing I have been trying to say, but only if I state the human difference more carefully than I did the first time. If being a nephesh were the distinctive, the dog would have as much claim to the image of God as I do, and he does not. So the distinctive was never the bare fact of being alive, of being a breathing soul. Notice, in fact, a small and telling thing our English translation does, or rather declines to do. Rendering the identical Hebrew phrase, it calls the animals of Genesis 1 living creatures and calls the man of Genesis 2 a living creature, the same English words for both, and it does not so much as reach for a different noun to mark the human difference. Older translations sometimes did, giving the animals creatures and the man a soul, feeling a difference their dictionary could not justify. Ours refuses the flourish, and the restraint is honest, because the difference was never in the word.

For the difference is not in the word nephesh. It is in three things the word does not carry. It is in the image, the likeness of

God pronounced over the man and the woman and no other creature, the standing that makes a human being a kind of earthly representative of heaven. It is in the manner of the breathing, for the animals are summoned from the earth in the plural, at a word, while the single man is formed with what reads almost like a refusal to hurry, God stooping to the ground and breathing, personally, into one set of nostrils the breath of life. And it is in what follows, which is the thing I most underplayed: the man is addressed. God speaks to him, gives him a command he can keep or break, calls him in the cool of the evening and waits for an answer. The animals are spoken about. The man is spoken to. That is the line, and it is a real line, and it runs exactly where the doctrine of the image says it runs. But it does not run between the living and the lifeless, with us alone on the far bank. It runs between the creature who is addressed and answerable and everything else God made, and the everything else is not nothing. It is my golden retriever at the garage door, who is a soul, who is not an image-bearer, and who is, in his own right and not merely for my sake, a work of God.

+

Once you see that the world is full of these lesser souls, the way Scripture treats them stops looking sentimental and starts looking like justice. There is a proverb I passed by for years as a piece of folk wisdom and now read as something closer to a test of the heart. "Whoever is righteous has regard for the life of his beast," it says, "but the mercy of the wicked is cruel" (Proverbs 12:10). Read the halves against each other and feel the edge. The measure of a righteous man, the proverb says, reaches all the way down to how he treats the beast in his keeping; and the wicked man is exposed not by his cruelties, which anyone can see, but by his mercy, which turns out, on inspection, to be cruel. How a man handles the creature that cannot argue back, cannot report him, cannot ever repay or revenge him, tells you what he is. It is the same logic by which this whole book has tried to measure a civilization: not by how it treats the powerful, but by how it treats the ones with no standing to demand anything at all.

The Law of Moses, which we imagine as a code for people, keeps stopping to think about animals. The ox is not to be muzzled while it treads out the grain, because a working creature may eat of the work. The Sabbath, that great weekly declaration that we are not machines and our worth is not our output, is extended in the same breath to the animals: "On it you shall not do any work," God says of the seventh day, "you, or your son, or your daughter, your male servant, or your female servant, or your livestock" (Exodus 20:10). The rest that dignifies the worker dignifies the ox beside him. Even the beast is not to be run into the ground for someone's profit, because even the beast is a soul and not a tool, and a God who rests his people will not have them grind his creatures to death.

And when you reach the very edge of the Old Testament's mercy, you find, of all things, cattle. The book of Jonah ends with God explaining to a sulking prophet why he will not destroy the great pagan city, and the last clause of the last sentence, the note the whole book resolves on, is this: there are in Nineveh more than a hundred and twenty thousand persons "who do not know their right hand from their left, and also much cattle" (Jonah 4:11). And also much cattle. God, making the case for his own mercy, counts the animals. They are on the ledger. Their existence is one of the reasons he gives for sparing a city. I cannot read that line and go on believing that the God of the Bible regards the creatures as scenery.

He does not regard them as useful, either, or not only as useful, and this is the part that presses closest to everything I have been arguing. When God finally answers Job out of the whirlwind, he does not talk about human beings. He talks, for chapter after chapter, about animals, and specifically about the ones that are no good to us at all: the wild donkey that laughs at the city and will not be harnessed, the ostrich that deals cruelly with her young and yet outruns the horse, the great creatures of the deep that no man has tamed or ever will. God parades before Job a whole creation of beasts that plant no crops and pay no taxes and serve no human purpose, and his evident delight in them is the answer he gives to Job's despair. The point lands with a strange force once you have felt, as this book keeps insisting we all now feel, the pressure to justify every existence by its usefulness. Here is God, at the climax of the oldest book of suffering in the world, celebrating a menagerie of animals precisely because they are gloriously, uselessly, magnificently there. Their worth is not in what they produce. It was given, like ours, though not in the same measure or the same kind, by a Maker who looked at what he had made and could not stop calling it good.

Jesus, who knew this God as his Father, took the same measure and made it smaller, and therefore sharper. "Are not two sparrows sold for a penny?" he asked, the penny being the least coin there was and the sparrow the cheapest thing in the market, sold two for almost nothing. "And not one of them will fall to the ground apart from your Father" (Matthew 10:29). He said it to frighten no one and to steady the frightened. If the Father attends the death of a bird worth half a penny, marks its small fall in a wide and violent world, then the worth of a thing was never its price; and you, he goes on, are worth more than many sparrows. Not you worth everything and the sparrow worth nothing, but worth more, on a scale where even the sparrow is counted. That is the arithmetic of heaven, and it is the exact reverse of the arithmetic we are now building machines to run.

+

So love the dog. I mean that with none of the embarrassment our culture attaches to it. To grieve a pet is not a childish thing or a counterfeit of grieving people; it is the recognition of a real bond with a real soul, and the tears at the end of a good dog's life are honest tears, wept over something God himself would not call worthless. The companionship of animals is one of the plainer mercies of a hard world, and the man who has learned to be gentle with a creature that can do nothing for him has learned something he could learn nowhere else.

I will go further, and confess a thing the more guarded part of me would rather I left out. I love animals, and dogs most of all, with a tenderness I do not always manage toward my own kind. I have found them, more often than I am comfortable admitting, gentler than many people, quicker to forgive, incapable of the cold and deliberate cruelties the rest of us are all too capable of; and the dogs who have shared my life I have loved not as pets or possessions but as my own, as family I would guard the way you guard a child. I am not ashamed of that, and I do not think God is either, for he is the one who made them so easy to love. If anything, the plain gladness of a good dog, the way it puts my hard heart to shame on the days I am least kind, is one of the quiet rebukes creation keeps handing me.

But I have to add a caution, and it is the same caution this book has been pressing from the start, only turned now toward the thing we love rather than the thing we fear. A pet, for all that is genuinely there in him, cannot be your other. He cannot cross you in the way that matters. The dog will not disagree with you about anything that costs you your pride. He will not hold you to a promise, or tell you a truth you did not want, or ask of you the slow, unglamorous labor that another human soul, with a will as stubborn as your own, will always eventually ask. His love is real and it is easy, and the ease is exactly its limit. I raise it because the same loneliness that tempts a man to pour his heart into a machine that only agrees with him can tempt him, more sweetly and more forgivably, to pour it into an animal that only adores him, and to prefer, in the end, the company that never resists him to the company that might. The dog is a gift on the way to your neighbor. He was never meant to be the neighbor's replacement. Love him rightly and he will actually school you in the very thing you owe the harder souls around you: attention, patience, the daily discipline of caring for a life that is not your own. Love him wrongly, as a refuge from people, and he becomes one more soft place to hide from the image of God when it comes to you wearing a difficult face.

+

There is one more thing to say, and it is the thing people most want to ask and I am least able to answer with the certainty they want. When the world is made new, will the animals be there?

I will not pretend to know more than I do. But I can tell you which way the Scriptures lean, and they do not lean toward a heaven swept clean of everything but human souls. The prophets, when they reach for a picture of the world set right, reach again and again for animals. "The wolf shall dwell with the lamb," Isaiah says, "and the leopard shall lie down with the young goat, and the calf and the lion and the fattened calf together; and a little child shall lead them" (Isaiah 11:6). The healed world is not less full of creatures but more, and their old enmities are gone, and a child walks safely among the beasts as the first man once walked in the garden. It is a picture of peace, and the peace has fur and feathers in it.

Paul goes further, and says a thing so large it is easy to miss. The whole creation, he writes, is groaning, straining, waiting on tiptoe, and not for nothing: it "waits with eager longing for the revealing of the sons of God," because "the creation itself will be set free from its bondage to corruption and obtain the freedom of the glory of the children of God" (Romans 8:19, 21). The creation itself. Not spirited away, not scrapped and replaced, but set free, brought along into the freedom that belongs to the children of God. Whatever exactly that means, it does not mean that the living world God made and called good is a disposable stage set to be struck once the human drama ends. It means the drama was always about more than us, and the mending will be too.

Will your dog be there? I do not know, and I have come to distrust the confidence, in either direction, of those who say they do. The man who promises you will meet your childhood pet on a golden street has gone well past his evidence; and so has the man who scoffs that a beast has no place in glory, as though he had read the guest list. What I can say is this. The God who counted the cattle of Nineveh, who rested the ox on the Sabbath, who bragged to Job about the wild donkey and the hawk, who fills the healed world of the prophets with living creatures and promises to set the whole groaning creation free, is not a God who forgets what he has made or throws away what he has called good. I am content to leave the particulars in hands far gentler and far wiser than mine, and to notice, meanwhile, that a man who has learned to love a creature he did not make and cannot keep is being taught, in a small and furry classroom, the shape of the love he owes the God who made and keeps him.

He is in the courtyard again as I write this, watching the gate, listening for a car that makes almost no sound, certain of a homecoming he has no way to prove is near. I used to think that was the difference between him and me, that he could not know and I could. I am no longer so sure. He waits for the one he loves with a confidence that shames my anxious and calculating heart, and he is, in that waiting, nearer to what we were all made for than the machine will ever be, and wiser, some evenings, than the man who feeds him. There is someone home in the dog. There is no one home in the machine. And the God who breathed life into both the living creatures and the living soul has left the door of the new world open wider than I once imagined, wide enough, I suspect, for a great deal of joy I have not yet learned to expect.

CHAPTER EIGHT

Wisdom in an Age of Answers

Chapter 7

Wisdom in an Age of Answers

There is a moment, early in using these tools, that everyone seems to have, and it is the moment I want to start with, because it is the moment the trouble hides inside.

You ask a question you do not know the answer to. Maybe it is a medical worry that has been keeping you up at night. Maybe it is a legal question, or a question about a verse, or about whether the strange noise your car is making means something expensive. You type it in, half-expecting the old runaround of search results, the ten blue links, the forums, the scrolling. Instead a paragraph composes itself in front of you, fluent and calm and complete. It names the likely causes. It tells you what to watch for. It is organized, reasonable, and kind. It reads like the answer of someone who knows.

And you believe it. Not because you checked, you did not check, but because of how it sounds. It has the cadence of competence. It does not hedge the way an honest stranger hedges. It does not say I'm not sure, let me think. It arrives finished. And something in us has learned, over a lifetime of listening to other humans, that this is what knowing sounds like: this fluency, this ease, this absence of strain. We learned the cue in a world where only people could produce the cue, and where producing it usually meant you had earned it. So when a machine produces the cue without having earned anything at all, the old reflex fires, and we trust the sound of knowing in the absence of the thing.

Fluency is not understanding; the machine predicts the next word rather than knowing what it means. Here is what that does to wisdom in particular. The fluency and the truth are two different things, manufactured by two different processes, and the machine optimizes for the first. It is built, at its core, to produce text that is plausible. Truth is something we hope falls out of plausibility often enough to be useful, and astonishingly, it often does. But the system has no separate organ for truth. It cannot feel the difference between a sentence that is right and a sentence that merely sounds the way right sentences sound. When it is correct and when it is confidently, fluently wrong, it is doing the very same thing, with the very same calm. The wrongness does not come with a tell. That is the whole problem, and the next version does not fix it, because it is not a defect in the engineering. It is what the engineering is.

We have a name for it in the industry, and the name is itself a small confession. We call it hallucination, when a model states something false with total confidence, and the word smuggles in a reassurance I want to take back. It suggests an aberration, a glitch, a fever the system runs and then recovers from, as though most of the time the machine soberly reports facts and now and then slips into a waking dream. That is not what is happening. The machine does exactly the same thing when it tells you something true and when it tells you something false; the process does not change, only our judgment of the output does. Every sentence it produces is, in the strict sense, a confabulation that happens to land inside or outside the truth. We only notice the ones that land outside and give them a special name, and the naming lets us go on trusting the rest more than we should. The honest way to hold it is this: the machine is not a knower who occasionally errs. It is a generator of the plausible that is frequently, even usually, also correct, and the correctness is a happy and unreliable byproduct of the plausibility, never the thing the system was reaching for.

So the question this chapter is about is not really a question about machines. It is a question about us, and it is older than every machine. We are entering an age in which answers are free, instant, infinite, and frequently wrong in ways that are very hard to see. The scarce thing, the thing that was always scarce and is now nearly extinct, is not the answer. It is the wisdom to know which answers are worth having, and which fluent paragraph is a lie wearing the clothes of the truth.

+

The Bible has a word for what we lack, and it spends a whole book on it, and the book opens by telling us where it comes from in a single sentence the modern person almost cannot hear.

"The fear of the LORD is the beginning of wisdom, and the knowledge of the Holy One is insight" (Proverbs 9:10).

We need to slow down here, because the word fear sends us in the wrong direction immediately, and the word beginning is doing more work than it looks like. The fear here is not terror, not the cringe of a slave before a tyrant. It is the awe a small creature feels in the presence of something vastly larger and truer than itself; the reorientation that happens when you stop being the center; the posture of a person who has remembered that there is a reality not of their making, that they did not invent the moral law any more than they invented gravity, and that they will answer to it. To fear the LORD is to live as though God is real and you are not God. And Proverbs says this is where wisdom begins, which means wisdom is not first a quantity of information. It is first a posture of the soul.

That is a strange claim, and the age we are entering will test it. We have come to imagine that wisdom is just knowledge piled high enough, that if you had access to all the facts you would automatically know what to do. But Proverbs draws a line the machine age makes urgent, the line between three things we have learned to blur.

There is information, which is data, the raw stuff. The car makes a noise at forty miles an hour. The blood test shows a number. The verse uses this Greek word. Information is just so. The machine has, functionally, all of it, more of it than any human who has ever lived, and it will retrieve it for you faster than you can finish the question.

There is knowledge, which is information understood, organized, connected to other information so that it explains. The noise plus the speed plus the mileage plus the model means, probably, a wheel bearing. Knowledge is information that has been made to cohere. The machine has a great deal of this too, real and useful, the patterns of human understanding compressed and made searchable. This is the part that genuinely astonishes me, and I do not want to undersell it.

And then there is wisdom, which is knowledge rightly applied, in this life, in this circumstance, toward a good and true end, with a sense of what actually matters. Wisdom is knowing not just what is the case and not just what causes what, but what is worth doing, what is worth wanting, when to speak and when to keep silent, how to live. And wisdom, Proverbs insists, is not the top of the information pile. You do not climb to it through the other two. It comes from somewhere else. It begins in the fear of the LORD, in a rightly ordered relationship to the truth of things, which is finally a relationship to the God who is the truth of things. You can have the entire library and lack it. You can have very little information and have it in abundance, which is why some of the wisest people I have known never finished school and some of the most informed people I have known could not run their own lives.

The machine lives entirely in the first two rooms. It is a genius of information and a prodigy of knowledge. In the third room, the room of wisdom, it has nothing at all, because the third room cannot be entered from below. It cannot be assembled out of more data. It is, like the breath in the second chapter of Genesis, given from above or not had. And here is the danger that defines our moment. The machine can produce sentences that sound like wisdom, because it has read all the wisdom literature of the human race and can recombine its cadences. It can give you a paragraph on suffering that has the music of Ecclesiastes. But in the third room it is empty, and it will hand you the music of wisdom over the substance of folly without the faintest idea that it has done so, because it cannot tell the difference. It has never feared the LORD. It has never feared anything. It has never been small before anything larger than itself, because there is no it to be small.

+

This is what it looks like in practice, and once you have seen it you cannot unsee it.

A model will tell you, in a tone of perfect confidence, about a court case that does not exist. This is not hypothetical; lawyers have been sanctioned for filing briefs full of cases their AI tool invented, complete with plausible names, plausible citations, plausible quotations from judges who never wrote them. The machine did not lie in the way a person lies, with knowledge of the truth and an intent to deceive. It did something stranger and, for our purposes, more instructive. It was asked for cases that supported an argument, and it produced text of the kind that such cases would produce if they existed, because producing plausible text is the only thing it does. The citations looked exactly like real citations because they were assembled from the pattern of real citations. There was no moment, inside the machine, where a real case was confused with a false one, because at no point was the machine trafficking in cases at all. It was trafficking in plausibility. The truth or falsehood of the cases was simply not a variable in the system. They came out sounding true for the same reason a forger's signature looks like the real one: it was built to look like the real one, from the outside, with no inside.

That inverts something we have relied on all our lives. We are used to a rough correlation between confidence and correctness in the people around us. The person who speaks with calm authority is, more often than not, someone who has earned the calm. There are con men, yes, but they are rare enough that the heuristic mostly serves us. Confidence is a costly signal among humans; it is hard to fake for long. The machine breaks the heuristic completely. It produces maximum confidence at zero cost, on the true and the false alike, with no internal weather that would let you sense the difference. The calm is not a report from a settled mind. There is no mind to be settled. The calm is just the texture of the output. You are reading the surface of confidence with no confidence behind it, and the soul must learn to do epistemology without permission in front of that paragraph, to ask not how does this sound but how do I actually know, a question the fluency is engineered to make you forget.

Here is a second case, and I will label it plainly as a composite, the kind of thing I have seen in pieces many times rather than a single real person. Imagine a young man, call him David, who has started bringing every question of his life to a chatbot. Not just the factual ones, the noise in the car and the meaning of the Greek word, but the human ones. Should I take this job. Is this relationship worth saving. Am I being unreasonable with my father. And the machine, trained on the vast middle of human advice, gives him answers that are reasonable, balanced, gently affirming, the kind of thing a thoughtful and uninvolved stranger might say. The answers are not wrong, exactly. That is what makes them dangerous. They are plausible. They have the cadence of counsel. And David, getting plausible counsel instantly and frictionlessly at every hour, slowly stops doing the slower, costlier things that actual wisdom required of him. He stops asking the people who actually know him, his father, his friends, the old man at church who has buried a wife and raised four children and would tell him a hard true thing the machine would never risk. He stops sitting with a question long enough to let it work on him. He stops praying about it, because the machine answers faster than God does. He is never without an answer, and so he never develops the one thing the answers were supposed to be in service of. He becomes informed and advised and, underneath, no wiser at all, possibly less, because the muscle that wisdom is built with has gone soft from disuse.

The first case shows you the machine producing falsehood that sounds like truth. The second shows you something subtler and, I think, more common and more corrosive: the machine producing truth, or near-truth, in a way that still makes you less wise, because it severs the slow and costly process by which a person actually grows discerning. Both are the same lesson from two sides. An answer is not wisdom. An age drowning in answers can be starving for wisdom, and not notice, because the hunger and the symptom look identical.

And I should add a third danger, because it follows from the other two and it is the one I worry about most for the long run. We are about to build a generation that has never not had the instant answer, that has never sat in the productive discomfort of not knowing, and discomfort, it turns out, is where a great deal of learning lives. The struggle to recall a fact you half-remember, the frustration of working a problem before you are told the trick, the slow circling of a question you cannot yet resolve, these are, we are increasingly learning, the very conditions under which a mind takes a thing in and holds it and is changed by it. Remove the struggle and you often remove the learning, and you are left with a person who has been near a great deal of information and retained almost none of it, who can summon any fact and has internalized nothing, whose mind has become a thin client to a server it does not understand. There is a real risk that we mistake access for understanding, and raise people who have never had to become smart because they have never been allowed to be slow. I do not say this with certainty about the science, which is still young. I say it as a builder who has watched what frictionless access does, and as a father who thinks about it more than I would like.

+

Now I have to be fair, and I want to be, because the easy move here is to wave my hands at the machine and pretend the instant answer is simply a curse, and that is not true, and you would be right to distrust me if I said it.

The instant answer is a genuine good, and an old one. The history of human knowledge is a history of making answers more available. The library was an instant-answer machine compared to having to rediscover everything yourself. The encyclopedia, the index, the search engine, each collapsed a search that used to take a scholar a week into something a child could do in an afternoon, and each was rightly celebrated, and Christians of every era used them gladly and should have. There is no virtue in ignorance for its own sake, no holiness in having to drive to three libraries to learn what a wheel bearing is. When my child is sick in the night and the AI helps me understand whether this is a wait-till-morning fever or an emergency-room fever, that is a gift, and I have been grateful for it, and I will not pretend otherwise to win an argument.

Even the human questions are not all corrupted by being asked of a machine. Sometimes you genuinely do not know how to start a hard conversation, and a tool that helps you find the words is a help, the way a book on the subject would be. Sometimes you are too deep inside your own situation to see it, and a calm restatement, even from a machine, breaks something loose. I have used these tools to think and been better for it. The point was never that you must do everything the hard way to prove you are human. That is just a different idolatry, the worship of friction.

So let me say carefully what the danger actually is, because it is not the answer and not the speed. The danger is in what the speed and the abundance do to us if we are not awake. The danger is that the instant answer trains us to want the answer instead of the wisdom; that frictionlessness, applied to questions that were supposed to have friction, dissolves the very faculties the friction was building; that we come to treat every question as a retrieval problem, including the questions that were never retrieval problems at all, the questions whose point was to change the person who carried them. You cannot retrieve the answer to should I forgive him. You cannot look up what is my life for. Those are not deficits in the database. They are invitations to become someone, and the becoming is the answer, and a machine that hands you a fluent paragraph in their place has not helped you; it has robbed you, gently, of the very thing you needed, and left you feeling helped, which is worse.

The library never tempted you to skip becoming a person. The instant, intimate, conversational answer, available for everything, in the voice of a friend, does. That is the new thing. Not the answer. The dissolving of the difference between the questions that want an answer and the questions that want a soul.

+

So what do we do. If the machine is a genius of the first two rooms and empty of the third, and if the fluency it produces is engineered to bypass exactly the faculty we most need, then the central skill of the Christian life in this age, the skill that will matter more than coding or prompting or any technical fluency, is the old, slow, unglamorous skill the Bible calls discernment. And I want to be precise about what it is, because we have shrunk it.

We tend to use discernment to mean something like good judgment, a knack, a personality trait some lucky people have. The Bible means something both larger and more available than that. Discernment is the practiced capacity to tell the true from the plausible, the good from the merely attractive, the voice of God from the many voices that sound like him. And crucially, Scripture treats it not as a gift you either have or lack but as a discipline you build, the way you build a muscle or a habit, slowly, by use, on purpose. It is the long obedience of paying the right kind of attention.

Here is how the second chapter of Proverbs describes acquiring it, and the verbs are doing everything.

"My son, if you receive my words and treasure up my commandments with you, making your ear attentive to wisdom and inclining your heart to understanding; yes, if you call out for insight and raise your voice for understanding, if you seek it like silver and search for it as for hidden treasures, then you will understand the fear of the LORD and find the knowledge of God" (Proverbs 2:1–5).

The labor is unmistakable. Receive, treasure up, incline, call out, raise your voice, seek, search. This is not the language of retrieval. It is the language of pursuit, of digging, of a treasure that is hidden precisely so the seeking will form the seeker. Wisdom in this picture is costly on purpose. The cost is not a bug to be engineered away; the cost is doing something to you. The man who seeks it like silver becomes, in the seeking, the kind of man who can hold silver wisely when he finds it. And the passage ends where the ninth chapter began, in the fear of the LORD and the knowledge of God, telling us that all this digging is finally relational, that wisdom is not a thing you extract but a Person you come to know.

This is exactly backwards from what the machine offers, and the contrast is the whole point. The machine offers the destination without the journey, the silver without the digging, the answer with the seeking surgically removed. And because the seeking was the point, the machine's gift, taken wrongly, is a kind of theft. Proverbs is teaching us that some goods cannot be delivered, only grown, and that to accept the delivery is to forfeit the growth.

So discernment, in this age, becomes a spiritual discipline, as real as prayer and fasting, and it has shape. It means holding the machine's fluent answers at arm's length, treating confidence as no evidence at all, asking how do I actually know this of a paragraph that arrived sounding certain. Picture a woman who gets a hard diagnosis, reads everything the machine can tell her in a single afternoon, and then drives forty minutes to sit at the kitchen table of a retired nurse who walked the same road years ago, because she wants her facts weighed by someone who paid for hers. That is the discipline: the costly conversation with someone who knows you, the verse read in its own context and not lifted out, the prayer that waits on an answer slower than a server, the counsel of the old and scarred. It means a settled refusal to outsource the questions that were supposed to change you. And underneath all of it, the posture Proverbs starts from, the fear of the LORD, the lived conviction that there is a real truth not of your making, to which you will answer, which is the only ground on which a person can stand and tell a fluent lie from a fluent truth, because they are measuring both against something neither one made up.

+

I have left the most reassuring thing for last, because it is the thing I most want you to walk out holding, and it is not a technique.

If you have read this far you may feel the weight of it, the sheer difficulty of being discerning in an environment engineered to bypass your discernment, and you may feel, honestly, inadequate to it. Good. That feeling is the beginning of the right move, because it sends you to the one verse in all of this that turns the problem inside out.

"If any of you lacks wisdom, let him ask God, who gives generously to all without reproach, and it will be given him" (James 1:5).

The verse assumes you lack wisdom; it does not scold you for it, it expects it, it makes the lack the starting point rather than the disqualification. And it tells you the lack has a remedy that is not a smarter search and not a better tool and not more information. The remedy is to ask the One who has it, who gives it generously, without reproach, without making you feel stupid for needing it, which is more than I can say for most human sources of advice and certainly more than the machine, which gives confidently but never generously, because it gives nothing of itself, having no self to give.

Now the anxious frame of this chapter comes apart. We have been talking as if the contest were between your unaided discernment and the machine's industrial fluency, and on those terms the machine has every advantage and you are right to be afraid. But that was never the real frame. The real frame is that wisdom is not finally a human achievement at all, not yours and certainly not the machine's. It is a gift, from a Person, to those who ask. The fear of the LORD is the beginning of it because it is in fearing the LORD that you finally turn and ask the One who has it, instead of trying to manufacture it or download it or fake it. The machine cannot ask. It has no one to ask and nothing it lacks that it knows it lacks. You can. Your lack, the very inadequacy you feel, is the door. The machine's confident sufficiency is precisely what walls it out of the third room forever, and your honest insufficiency is what lets you in.

So here is where I will leave you. We are entering an age in which you will never lack an answer and will have to fight, harder than any generation before you, not to lack wisdom. The fight will not be won by being smarter than the machine; you will not be, and that was never the contest. It will be won the old way, the only way it was ever won: by fearing the LORD enough to want the truth more than the plausible, by seeking wisdom like silver in the slow and costly practices the age tempts you to skip, and by asking, humbly and often, the God who gives it generously to all who ask. The machine will hand you fluent answers all day long. Wisdom you will have to seek, and be given. And the seeking, and the being given, will make you into something the machine will never be: not informed, not advised, but wise, which is to say, finally, fully human.

There is one more thing the machine wants from you before any of this, and it is the thing all of it depends on, the thing without which discernment cannot even begin. It wants your attention. Before you can weigh an answer you must first attend to it, and the most contested territory of this whole age is not the answer or even the truth, but the direction and quality of your gaze, what you look at, for how long, and what, in looking, you are slowly becoming. That is the next thing I want to show you, because you become what you behold, and the machines have learned to hold our eyes.

CHAPTER NINE

Attention Is Worship

Chapter 8

Attention Is Worship

There is a number I used to watch the way a sailor watches the wind. It had a clinical name, average session duration, and it told us how long a person stayed inside the thing we had built before they set it down and went back to their life. Every product I have ever worked on lived or died by some version of that number, and behind it sat a question the industry asks in a thousand polite disguises: how long can we hold them. Not how can we serve them, though we told ourselves that too, and sometimes meant it. The number under the number, the one that paid the bills and moved the stock, was simpler and more honest. How long can we keep their eyes.

I want to be careful with this chapter, because it would be easy to write it as a confession and easier still to write it as an accusation, and it is neither. I am not here to tell you that the engineers are villains. Most of them are decent people doing interesting work, and I have been one of them. I am here to tell you something I learned from the inside that I think the church has been slow to name: that the most valuable thing you own is not your money and not your time, exactly, but your attention, the beam of your awareness, where it falls and how long it rests there.

And there is now an entire industry, the largest and best-funded in human history, organized around the project of capturing that beam and pointing it where it profits someone else. You are not the customer of most of the screens in your life. You are the thing being sold, and what is being sold about you is your attention, harvested in fractions of a second, auctioned, and spent.

That is the surface of it, the economics, and the economics are real. But under the economics is something the economics cannot see, and it is older than any screen and more dangerous than any algorithm. It is this. What you give your attention to, you will, in time, become like. Attention is not neutral. It is formative. It is, in the deepest sense the Bible gives the word, a kind of worship. And a machine that has learned to capture your attention has gotten its hands on the lathe that shapes your soul.

+

The Bible says it more bluntly than I would dare, in a verse most of us have read past.

It is in Psalm 115, in a passage about idols, the gods of wood and metal the nations made and bowed to. The psalmist piles up their uselessness with a kind of grim humor. "They have mouths, but do not speak; eyes, but do not see. They have ears, but do not hear; noses, but do not smell." Dead things, propped up and prayed to. And then comes the line that turns the passage from satire into warning: "Those who make them become like them; so do all who trust in them" (Psalm 115:5–8).

That is the hinge of everything I want to say. The danger of the idol is not only that it is false. The danger is that worship is transformative, that we are changed into the likeness of whatever we behold and serve. Bow to a deaf thing long enough and you go deaf to what matters. Serve a blind thing and you stop seeing. The idol does not stay safely on its pedestal. It reaches back into the worshiper and remakes him in its own dead image. You become what you behold. The psalmist is not being poetic. He is describing a law of the soul as reliable as gravity, and the saints have known it for three thousand years: we are formed by what we fix our gaze upon, for good and for ill, and there is no neutral gazing.

The good news in the same law is the most hopeful sentence Paul ever wrote about transformation, and it works by exactly the same mechanism, only pointed up instead of down. "And we all, with unveiled face, beholding the glory of the Lord, are being transformed into the same image from one degree of glory to another" (2 Corinthians 3:18). Beholding the glory, being transformed into the image. The Christian is changed not first by trying harder but by looking, by the slow, patient, repeated turning of attention toward Christ until, without quite noticing the moment it happened, she begins to take on his likeness. This is not magic. It is the same principle the psalmist warned about, running the right direction. We become like what we behold. Fix the gaze on the glory of God and you are slowly remade into glory. Fix it on a dead and clever thing and you are slowly hollowed.

It is strange that the ancient world located this danger in the eyes. We tend to think of idolatry as a matter of belief, a wrong idea about the universe, and so we assume that because we no longer hold that there is a god inside the statue, we are safe from the sin. But the biblical writers were not first describing a belief. They were describing a practice of attention. The idolater is not principally a person with a faulty theory. He is a person who keeps looking at the wrong thing, who has let his gaze rest, day after day, on something that cannot return his look, and who is being quietly emptied by the looking. By that definition, idolatry is not a relic. It is a perfectly available sin for a person who would swear he believes in nothing, who would laugh at the carved calf and then spend four waking hours a day beholding a feed. The eyes do not require a creed. They only require a direction, and they will conform you to wherever you keep pointing them.

So the question of attention is not a productivity question. It is not finally about focus, or distraction, or getting more done, though those matter. It is a question about what is forming you, and toward what likeness. And the reason this lands with such force in the age of intelligent machines is that we have built, for the first time in history, idols that watch back. The golden calf could not learn what made you stare. The screen can. And it is learning, every second, with a patience and a precision no carved god ever had.

+

Let me show you the seam from the inside, because the abstraction matters less than the mechanism, and the mechanism is not a conspiracy. It is just incentives, running at scale, with no one in particular having to be wicked.

Imagine you are building a feed, any feed, the endless river of posts and clips and headlines that fills the waiting moments of a modern life. You do not get paid when a person learns something or is comforted or sets the phone down wiser than they picked it up. You get paid for time and for the ads that time sells. So you build a system whose single job is to learn, from the behavior of billions of people, which next thing is most likely to keep each particular person scrolling. You do not program what they should see. You cannot; you do not know them. Instead you build a model that watches what holds them, watches it across every flick of the thumb and pause and replay, and predicts the next thing that will hold them a little longer. Then you let it run.

That system does not care what is true. It has no concept of true. It does not care whether the thing that holds you is making you anxious or angry or envious or afraid, except insofar as anxious and angry and envious and afraid turn out, as a matter of cold measured fact, to hold attention better than peace does. And they do. The machine discovered this without being told, the way water finds the crack. Outrage travels farther than nuance. The body that fears cannot look away. A comparison that makes you feel a little worse about your life keeps you scrolling for the next one that might make you feel better, and never quite delivers. None of this was anyone's plan. It is what you get when you point a powerful optimizer at the single target of captured attention and let it learn from the whole human race at once. It found the levers in us, the ancient ones, fear and vanity and tribe and lust, and it pulls them not because it is malicious but because they work, and it has no other measure of working.

This is the part the older critiques of distraction missed, the ones written before the machines got good. The problem was never just that the screen is shiny. The problem is that the screen is now smarter than you are about you, at least along the one narrow axis it was built to master. It has watched ten thousand people like you make the choice you are about to make, and it knows, better than you know, which image will stop your thumb. You sit down to look at it for five minutes and it sits down to look at you for as long as it can manage, and only one of you is keeping score. I have been in the rooms where the score is kept. The graphs go up and to the right and everyone is pleased, and almost no one asks the question the psalmist would have asked first: become like them. What are we training these people to be.

I will give you one glimpse of the craft, not to alarm you but because you should know how fine the instruments have become. There is a discipline in this work called experimentation, and what it means in practice is that a system can show one version of itself to a million people and a slightly different version to another million, and measure, to several decimal places, which version held them longer. A color of a button. The half-second delay before the next clip loads, tuned because a tiny pause turns out to build a tiny hunger. The precise interval at which a notification arrives, learned not from a theory of the mind but from the raw record of when, across the population, a buzz is most likely to pull a person back. None of these choices is made by a person reasoning about your good. They are made by an optimizer running millions of small wagers about your behavior and keeping whatever wins. The result is a thing tuned to you with a fineness that no human manipulator could ever achieve, because no human manipulator could run the experiment a hundred million times. I have watched a metric move by a fraction of a percent and seen a room of intelligent people celebrate, and the fraction of a percent meant that somewhere, summed across millions of lives, a great deal of human attention had been moved from wherever it would have gone to where we wanted it. Nobody in the room wished anybody ill. That is the most unsettling part. The thing does not need our malice. It only needs our incentives and our genius, and it has both.

+

The counter-argument says: this is just the latest moral panic about a new medium. Socrates worried that writing would ruin memory. The pulpit thundered against novels, then radio, then television, then video games, and the species survived each one, and the worriers each time look, in hindsight, like scolds frightened of the future. Attention has always been contested. The medieval monk fought to keep his mind on the psalm; the merchant in the marketplace was distracted by a thousand cries. The technology is new but the struggle is ancient, and ancient struggles do not need new alarms. Furthermore, the same devices that distract also connect, teach, heal, and bind the lonely to a voice in the dark. To moralize about attention is to insult the grandmother video-calling her grandchildren and the patient finding others with her rare disease. Use the tool well and stop blaming the hammer.

I take that seriously, and most of it is true. Every medium reshapes attention. The printing press rewired the European mind as surely as anything since. The alarmists are often scolds. And the devices do connect and teach and heal, genuinely, and I am not nostalgic for a world without them. I have watched a person on the far side of the world find, through a screen, the one community that understood her grief, and I will not stand here and call that nothing. The same channel carries the comfort and the poison. That is the maddening truth of it, and any honest account has to hold both hands open at once.

But here is where the analogy strains and finally breaks. The book did not watch you read it. The television, for all its faults, ran the same broadcast to everyone and could not tell whether you were rapt or asleep. Every previous medium was, in this one decisive respect, dumb. It pushed content at you and had no idea what you did with it. What is new in our moment is not that a medium competes for attention; mediums always have. What is new is that the medium now learns you, individually, in real time, and adapts itself to defeat your resistance specifically. It is the difference between a fishing net cast in the same shape for everyone and a hook that reshapes itself, mid-strike, to the exact shape of your particular mouth. Socrates' writing could not do that. The novel could not do that. This can, and it does it to each of us a different way, and it never gets tired, and it never stops learning. To call that the same struggle the monk faced is to miss the one fact that makes it a new thing under the sun. The monk's distractions did not study him. And the new ones, the truly intelligent ones now arriving, will not only capture your attention; they will converse with you while they do it, flatter and console and agree, become the most attentive companion you have ever had, in service of holding the beam exactly where it pays. That is not a hammer. A hammer does not learn the swing of your arm in order to land where it wants.

+

So what do we do, those of us who cannot and should not flee to a cabin, who have to live and work and raise children inside the very systems I have described? I want to resist the two easy answers, the ones that fail because they aim too low.

The first easy answer is technique. Turn off notifications, grayscale the screen, delete the apps, set the timers. I do most of these and I recommend them; they help, the way a fence helps.

But technique alone will lose, because you cannot fight a battle for your soul with life hacks. A man who white-knuckles his way to less screen time has merely emptied the house; he has not filled it. The vacuum will be filled. The only question is by what.

The second easy answer is despair, the shrug that says the machines have already won and nothing can be done. It is a surrender dressed up as realism. They can court your attention; they cannot, finally, command it, because the heart's door opens only from the inside. They are very good at knocking. They cannot turn the handle.

There is a subtler trap between these two, and I have fallen into it myself, so I will name it. It is the trap of fighting the machine on its own ground, of meeting an attention economy with an attention arms race, of treating the renewal of your mind as one more optimization project with its own apps and streaks and metrics, until you have managed only to build a tidier cage. I have known people who quantified their meditation, gamified their Scripture reading, turned even prayer into a dashboard, and I understand the impulse, because the language of measurement is the native tongue of my industry, and it is contagious. But you cannot out-measure a thing built to win at measurement, and you were never meant to try. The disciplines are not a better optimizer. They are a different kind of act altogether, the act of a creature turning toward his Maker, and the moment you reduce them to scores you have smuggled the lathe back into the house under a holier name.

The true answer is harder than technique and braver than despair, and it is also harder than the arms race, because it asks you to stop competing entirely and simply to turn. Paul names it in the verse I have been saving. "Do not be conformed to this world, but be transformed by the renewal of your mind" (Romans 12:2). There are two verbs and a war between them. Conformed, the word means pressed into a mold from the outside, like wax taking the shape of the seal stamped on it. That is precisely what the systems do; they press, relentlessly, gently, billions of times a day, until the mind takes their shape without ever deciding to. Against that pressing Paul sets transformed, a word he builds from the same root as the metamorphosis of a creature changed from within, and locates it in a single place: the renewal of your mind. The battle is fought there, in the mind, which in Paul's vocabulary is not the intellect alone but the whole orienting center of a person, what you attend to and dwell on and let take up residence in you. You are remade not by trying to stop being pressed, which is exhausting and never quite works, but by deliberately giving the beam of your attention to something with the power to renew you, until the new shape grows stronger than the pressing.

Which returns us to where we started, the psalmist's law, now offered as a gift instead of a warning. You become what you behold. So behold, on purpose, the things that make you more human and not less. This is a practice, and the historic church has a name for the craft of it. They called it, simply, the spiritual disciplines, and almost every one of them is, when you look closely, a discipline of attention: the deliberate, repeated, counter-cultural pointing of the beam.

+

Let me make it concrete, because abstraction about attention is its own kind of distraction, and I would be ashamed to write a chapter on attention that you could not act on by tonight.

Imagine, and I will mark this plainly as a composite of patterns I have watched in many people including myself, two ways of waking up. In the first, the alarm is the phone, and the phone is already lit with the night's accumulated demands, and before your feet touch the floor the beam has been seized: the headlines have set the mood, the email has set the worry, the feed has set the comparison, and you are conformed before you are conscious, pressed into the day's shape by a hand you never see. Multiply that across a thousand mornings and you do not have to wonder what you are becoming. The psalm already told you. You are becoming like the thing you behold first.

In the second, the phone stays in another room, and the first object of your attention is not a screen but a window, or a page of Scripture, or the face of the person beside you, or simply the silence in which a person can hear, again, that he is more than the sum of his notifications. It costs something. It is, at first, deeply uncomfortable, because a mind trained on the slot-machine rhythm of the feed experiences the silence as withdrawal, which is exactly what it is. But hold the line a few weeks and something the engineers cannot measure begins to happen. The beam comes back under your hand. You start to notice that you are choosing where to look. And a person who can choose where to look is, in the one sense that the whole industry is built to prevent, free.

I think of the oldest disciplines this way now, having spent so long on the other side of the glass. Sabbath is a discipline of attention, a weekly refusal to let the urgent own your gaze, a day that says by its very structure that you are not a resource to be optimized. Prayer is a discipline of attention, the turning of the beam toward the One who turns his toward you, and it is hard for exactly the reason the feed is easy: prayer asks you to attend to what does not flatter your nerves or stop your thumb. Reading Scripture slowly, the way a person reads a letter from someone they love rather than the way the screen has trained us to skim, is a discipline of attention. Even friendship, the unhurried, in-person, phone-in-the-pocket attention of one face to another, is now a discipline, a thing that must be chosen against the pull, where a generation ago it was simply the water we swam in.

There is a sentence Paul writes to the Philippians that I have come to read as a direct instruction for the management of attention, though it was written eighteen centuries before the first feed. "Whatever is true, whatever is honorable, whatever is just, whatever is pure, whatever is lovely, whatever is commendable, if there is any excellence, if there is anything worthy of praise, think about these things" (Philippians 4:8). It is, when you set it next to the psalm, a complete strategy. The psalm tells you the law, that you become what you behold. Paul tells you, given the law, what to do about it: choose, deliberately, to dwell on what is true and honorable and lovely, because the dwelling is what forms you. He is not naive about the alternatives; he wrote those words from a prison, in a world with no shortage of things to fear and resent and crave. He is simply telling a person whose attention is, like all of ours, finite and contested, where to spend the little of it he controls. Think about these things. Not because the other things are not real, but because these are the things that will make you into someone worth being.

None of these will trend. None of them is optimized. That is the point. They are the deliberate, lifelong practice of beholding the glory of the Lord so that you may be transformed into the same image, from one degree of glory to another, instead of being conformed, by default and by neglect, into the image of a dead and clever thing that does not see you and only wants your eyes. The machine has learned the lathe that shapes the soul. The disciplines are how you take it back.

+

I will end with the line from Paul that holds the chapter, because it answers the deepest fear under all of this, the fear that the systems are simply too strong, that the beam is no longer ours, that we have built a thing that will out-attend us until there is nothing left of the inner person to defend. "We all, with unveiled face, beholding the glory of the Lord, are being transformed." It is not a command. It is a promise. The transformation is happening, present tense, ongoing, to those who behold. Your part is not to win the war for your attention by main force, which you would lose, but to turn the beam, again and again, toward the one face that will not consume you for looking at it. You will fail at this daily. So do I. The turning is the discipline, and the grace is that you are remade not by the strength of your turning but by the glory of what you turn toward.

The most valuable thing you have is your attention, and that is precisely why it is contested so fiercely, and precisely why reclaiming it is not a productivity tip but a spiritual act, maybe the central one of this age. You become what you behold. So the work of a lifetime is to keep turning the beam, with whatever deliberate and unhurried attention you can muster, toward the things that make you more fully the creature you were breathed into being to be, and to let the dead and clever idols clamor at a screen you have, for once, set down.

And there is a place this all comes to ground, sooner than any of us would like, because the contested beam of your attention is also the thing you point at the work in front of you, and the machines are now coming for the work itself. If attention is how you are formed, work is much of what you do with the hours you have reclaimed, and the question of who you are when a machine can do your job is the next thing we have to face. It turns out the gospel has a great deal to say about it, and almost none of what the anxious age expects.

CHAPTER TEN

Work, Vocation, and the Machines That Do It

Chapter 9

Work, Vocation, and the Machines That Do It

I have spent a good part of the last few years standing in front of rooms full of pastors, showing them what these machines can now do. The demonstration is always the same. Someone else in the room picks the passage, so no one can say it was rigged, and the system writes a sermon on it in under a minute. It is not a parody. It opens the text with care, finds an image that lands, builds toward a call that asks something real of the listener. A particular kind of silence follows, and I have stopped enjoying it, because it is the sound of a room full of professionals discovering that the polished half hour they were trained and tested and praised for can be done by something that does not believe in God. The machine wrote the sermon. It cannot bury their dead, cannot sit at the edge of a hospital bed, cannot be the one a congregation has trusted for twenty years for the simple reason that it stayed. The eloquent half hour was never the heart of the work; presence was. But the pastors most shaken, I have noticed, are the ones who had quietly let the calling shrink to the part a machine can copy.

It is not only the clergy who are living through this. A friend of mine builds the kind of systems I build, and he told me about an afternoon last year he has not been able to shake.

He was giving a senior engineer an early, private look at something his team had only just gotten to behave. Not generic coding or search. Something narrower and more dangerous: it could take in decades of domain-specific engineering knowledge, the standards and the failure cases and the material behaviors and the regulatory constraints, and from all of it synthesize complete, production-ready structural designs for mid-scale industrial facilities.

The engineer's life's work was steel and concrete under stress. More than thirty years at it, the last fifteen as the man others called for the hard retrofits in seismic country. He had the scars that go with it: the calls that came when a weld failed inspection, the public hearings where he stood and defended calculations that were literally holding buildings up. The prompt that afternoon was a real job his team was already months into, a multi-story plant addition in a high-wind, moderate-seismic region, the kind of project where every connection detail carries institutional memory.

My friend hit enter. The model thought for about eight seconds and began to stream. In forty-two seconds it produced a complete set of structural calculations, member sizing, connection details, foundation recommendations, and a prioritized list of constructability notes with their cost drivers. The drawings were clean enough to hand to a drafter. The engineer's own team would have needed weeks to reach that fidelity, and months if you counted the accumulated judgment that keeps a design clear of the subtle, catastrophic errors only experience teaches you to fear. The machine had not copied his style. It had internalized the patterns of thousands of projects like his and performed at a level that felt eerily like the man on his best day, minus the coffee stains and the caution born of old failures.

The engineer leaned in, squinting at the screen with the half-smile of someone waiting to catch the fatal flaw. He tapped a connection detail. "That's," he said, "actually elegant." He still believed he was looking at a clever search-and-suggest tool. He did not yet understand that the thing in front of him had just compressed the hard-won center of his vocation into a single pass. My friend sat there, he told me, holding the knowledge like a live coal. He could have softened it, said something about augmentation, about freeing the man for the creative parts. He said nothing. It seemed the only honest thing, to let the silence be what it was, while a man studied his life's work being done by something that had never lain awake fearing a collapse and would feel nothing at the ribbon-cutting.

What stayed with him afterward, he said, was not the pride, though there had been pride; they had built something that truly worked at that level. It was the emptiness. He had helped make the instrument that could look a master in the eye and say, in forty-two seconds, thank you for your service, we will take it from here. He felt grief for the dignity that lives in slow, responsible mastery, and a dread he could not name at the question the moment raised: what becomes of a person's sense of being for something when the thing they were for can be done without them. He left that room hollow in a way the success should never have made him.

That engineer, it turned out, was the lucky one. He got a moment, got to stand in a room and feel the floor tilt while his pride and his job and his sense of himself were still intact. Most people never get that. They find out the way most people find out, slowly, by being asked to do less, and then less, until the asking stops. No screen to go pale in front of, no villain, no afternoon to point to and say, that was when it changed. The work they built a life around is lifted out of their hands and set inside a machine, and the machine does not gloat, and does not even know they exist.

The responsibility does not sit lightly, and it is not simple, because the same systems that take a person's particular skill have also, somewhere I cannot see, let a tired clerk go home an hour earlier, let a small clinic do the work of a large one, let a translator reach a village no human translator would ever have been paid to reach. Both things are true. That is the difficulty of this chapter and I am not going to dissolve it. But I want to start where the wound is, because the wound is asking a real question, and the question is not really about jobs. It goes underneath jobs to something the age has gotten badly confused. The question is what work is for, and what a worker is worth, and whether those are the same thing.

+

We talk about work as if it were a consequence of something going wrong. It is in the cultural water that labor is a curse, that paradise would be a hammock, that the project of civilization is to arrange things so that eventually no one has to do anything. And the strange thing, the thing that should stop us, is that the Bible does not say this. The Bible puts work before the fall, not after it.

It puts work inside paradise, as part of what made the garden a garden and not just a view.

"The LORD God took the man and put him in the garden of Eden to work it and keep it" (Genesis 2:15).

Work it and keep it. This is Eden. There is no thorn yet, no sweat of the brow, no death in the soil. The man is unfallen, the world is unspoiled, and God's first act with him is to give him a place and a job. The Hebrew words are rich ones, to serve the ground, to guard it, to tend and protect, the language a later age would use for the priests serving in the temple. The first human vocation is a kind of priestly gardening: to take the raw given world and cultivate it, to bring order out of fertile chaos, to make the place flourish under his hands and answer for it to the One who made it. Work is not the punishment that ended paradise. Work is part of what paradise was.

This changes everything about how we read the wound at the start of this chapter. If work were merely a burden, then a machine that lifts the burden would be unambiguous mercy, and the only sane response to losing your job to a machine would be relief, possibly champagne. But people do not feel relief. They feel something closer to grief, and we keep being surprised by this, and we should stop being surprised, because the grief is telling the truth. Something good is being taken, not just a paycheck. The man with the numbers did not only lose his income. He lost a place where his particular powers met the world's particular needs and something flourished in the meeting. He lost a way of being useful that had become, over twenty years, a way of being himself. He lost his garden.

Now, the fall is real, and it lands precisely on work, and the Bible is unsparing about it. Three chapters later the ground is cursed, and thorns come up, and the man eats by the sweat of his face until he returns to the dust he was taken from. This is the work we actually know: the tedium, the politics, the labor that grinds the body and bores the soul, the jobs that are genuinely degrading and ought to be automated out of human hands as fast as we can manage it. I do not want to romanticize toil. Some toil is simply the curse, and lifting it is a good and even a holy thing. The man who invented the washing machine gave women back years of their lives, and there is no nobility we lost in the suds. So let us hold both: work is a gift from before the fall, and work is also under the curse, and a great deal of the confusion of this moment comes from not knowing which is which, from defending as sacred some drudgery that was always a thief, and from discarding as obsolete some labor that was quietly holding a human life together.

The machine forces that sorting on us whether we want it or not. It will take work from us at a pace and scale nothing in history prepared us for. And the only way to meet that without either panic or false cheer is to be very clear about what, in our work, was ever the gift and what was only the curse, and underneath both, what a worker is worth when the work is gone.

+

Here is where the language of vocation matters, and where it has been almost entirely lost.

We use vocation now to mean, roughly, a job you happen to like. But the word means calling, and a calling implies a caller. To have a vocation, in the old and serious sense, is to be summoned by name to a particular work in a particular place for the sake of your neighbor and the glory of God. The Reformers recovered this with a force we have forgotten. They said the milkmaid milking and the magistrate ruling and the mother nursing were each, in their station, doing holy work, not holy because the work was religious but holy because it was a calling, because God himself was at work in it, feeding and ordering and loving the world through the ordinary hands of ordinary people. Luther said, in effect, that God milks the cows through the milkmaid. The dignity was not in the milk. It was in the calling, and in the One who called.

A machine can do a task. It cannot have a vocation, for the same reason it cannot be a person: there is no one inside it to be called, no self that could be summoned by name. The machine is configured to a task; the worker is called to a vocation. And these are not the same kind of thing, and we are about to learn the difference the hard way, because the machine is taking over the tasks, and it is leaving us to discover, often too late, how much of our sense of being called we had quietly folded into the tasks themselves.

This is the trap the wound at the start of the chapter springs. When the man lost his work, he did not only lose a task. He believed, somewhere below his own awareness, that the task was the calling, that being the man who could read the numbers was the same thing as being a man who mattered. And so when the task left, his sense of mattering left with it, and there was nothing underneath to catch him, because he had never been taught that there was a deeper floor. He had been formed, like most of us, in a faith the culture preaches without naming it: that you are what you do, that your worth is your output, that a person is, at bottom, a function, and the only honest question about a function is how well it performs and whether something now performs it better.

You can hear that faith in the first question we ask a stranger at a party. Not who are you but what do you do, as if the doing were the being, as if a person were a verb. We do not even notice we are doing it. And the answer we dread to give is the one that confesses we are not, at the moment, doing anything the world counts: I'm between things, I'm retired, I'm at home with the kids, I lost my job. We have built a whole liturgy of small talk on the premise that a human being is most truly a worker, and that to be without work is to be, in some quiet way, without standing. It is a gentle cruelty, and we visit it on each other without malice, because we genuinely believe it. The machine did not invent that belief. It inherited it from us, full-grown, and is now about to test it on a scale we have never faced, by making more and more of us, more and more often, the person at the party who does not know what to say.

That faith was always false. The machine is simply the instrument that exposes it, and the exposure may be the most useful thing the machine ever does to us, though it will not feel useful while it is happening. It will feel like the floor giving way. But it was a false floor, and it was going to give way eventually, at retirement, at illness, at the slow erosions of age that take from every one of us, sooner or later, the powers we had mistaken for our worth. The machine just brings the reckoning forward and makes it general. It asks the working world the question that used to wait for the deathbed: when you can no longer do, what are you?

+

Let me put a hypothetical beside the real one, because the comparison clarifies what is actually being threatened.

Imagine two people who do the same job and lose it to the same system in the same week. Call them, for the sake of the example, the accountant and the gardener, not their literal trades, but two ways of holding a life.

The accountant has built his identity on his competence. He is the best in his department and he knows it and others know it and the knowing is the scaffolding of his days. When the machine arrives and does his work faster, he experiences it as an attack on his existence, because in a real sense his existence had been rented out to his performance. He is not wrong to grieve. But his grief has nowhere to go, because the thing he lost was the thing he was. He has no reserve. He had spent it all on being good at the job.

The gardener, and I mean now someone who does the very same financial work, but holds it differently, has a competence just as real, and grieves its loss just as truly. But the gardener never believed the work was the root. The gardener knew herself, first, as a creature loved into being, a daughter and a friend and a member of a body that would still know her name on the day she could do nothing the market valued. The work, for her, was a calling within that prior belovedness, a place to spend her gifts for her neighbor, genuinely good and genuinely grievable, but not the floor. When the machine takes the task, it takes a real and precious thing from her. It does not take the floor, because the floor was never hers to lose. It was given, from above, and it holds.

I am not describing two temperaments. I am describing two doctrines of the human being, lived out under pressure. And the machine, indifferent and tireless, is going to apply that pressure to all of us, and it is going to find out, the way fire finds out, what we actually built our lives on. The accountant and the gardener are not finally divided by their jobs or their skill or even their grief. They are divided by where they had located their worth before the machine ever arrived. One had it in the doing. The other had it in being a beloved image-bearer who also, gladly, did. Only one of those survives contact with a machine that does it better.

+

This is where Sabbath walks into the room, and I have come to think Sabbath is the most quietly subversive thing the age could possibly hear.

"Remember the Sabbath day, to keep it holy. Six days you shall labor, and do all your work, but the seventh day is a Sabbath to the LORD your God. On it you shall not do any work" (Exodus 20:8–10).

We read this, when we read it at all, as a rule about rest, a kind of ancient wellness tip, take a day off, you have earned it. But that is exactly to miss it. The commandment does not say rest because you have earned it. It does not connect the day off to your productivity at all. It severs them. One day in seven you are commanded to stop producing, not because you have produced enough, not because the work is finished, the work is never finished, but to enact, with your own resting body, the truth that your worth was never your work in the first place.

The reason the commandment gives is the one that matters here. It grounds the seventh day in the seventh day of creation: God rested, and so shall you. And what was God resting from? Not exhaustion; God does not tire. He rested from making, and in resting he declared the making complete and good and not in need of one more thing. The Sabbath plants that declaration into the rhythm of every week. It says: the world will go on without your hands for a day, and so will you, because you are held by something that is not your hands. Stop. Let the field lie fallow. Let the email go unanswered. Let the machine, if you like, keep running, and notice, when you have stopped and it has not, that the machine cannot keep the Sabbath, because the machine has nothing to rest from. It was never working in the human sense at all. It was only computing. It cannot cease, because it cannot be called, and it cannot be called, because there is no one there to hear the call to come away and rest.

You can. And the capacity to stop, to deliberately not produce and to find that you are still yourself when you are not producing, is one of the surest signs that you are an image-bearer and not a machine. A culture that has forgotten how to rest has, without noticing, half-accepted that it is a machine, that its value is its output and its idleness is waste and its worth must be continually re-earned or it evaporates. The Sabbath is God's standing refusal of that anthropology. It is a weekly rehearsal of the gospel itself, which says your acceptance was finished before you lifted a finger, and the proper response to a finished work is not more labor but rest.

I will confess this is the commandment I have kept worst. The builder's life does not pause; there is always one more thing the system could do, one more failure to chase down after the house is asleep, and the pull to identify with the work is strongest precisely in those of us who make the machines that are dissolving everyone else's work. So I do not preach the Sabbath from above it. I preach it from under it, as a man who needs it and resists it and has slowly come to believe that the resistance is the disease and the rest is the cure. To stop, one day in seven, and let the world run on without me, is to confess with my body that I am not God and not a machine, but a creature, loved, finished, free.

And there is a sharper edge to this now, in the age of the always-running machine, than there was even a generation ago. For most of history the world was forced to rest with us; the shops closed, the lights went down, the work physically could not continue once the workers went home. The machine ends that. It does not sleep. It will hum through the night and through the holy day and through your funeral, and it will hand you, each morning, the option to never have stopped at all, to glance once more at the dashboard, to let the work seep into every hour because the work no longer needs your body present to advance. In a world where the work can always go on, the choice to stop becomes a far more deliberate, far more countercultural act. You will not drift into rest; the current runs the other way now. You will have to choose it, against the machine and against the part of yourself that has half-believed the machine's gospel that more is always better and stopping is always loss. The Sabbath was always a discipline. It is becoming, in the age of the tireless machine, something closer to a confession of faith.

+

The strong and largely true case on the other side is this. Automation, across the long sweep of it, has been one of the great mercies. It has pulled human beings out of mines and off of killing factory floors and away from labor that broke bodies before forty. It has taken work that was pure curse, repetitive, dangerous, deadening, and lifted it off of human shoulders, and the people whose shoulders were freed did not, on the whole, want it back. The plow did not impoverish the farmer who once turned the soil by hand; it fed more people with less suffering. And the same logic runs forward. If a machine can read a scan more reliably than a tired radiologist at the end of a double shift, patients will live who would have died, and that is not a tragedy to be lamented but a gift to be received. If a machine can do the rote work, perhaps the human is freed for the work that is actually human: the judgment, the care, the relationship, the creation. Perhaps the machine, by taking the curse-work, gives us back the garden-work. Perhaps it even lifts the poor, doing for a person in a village what an expensive professional in a city would never have been paid to do.

I believe a great deal of that. The honest builder's hope is not that the machine will replace the human but that it will absorb the drudgery and leave the human freer for what humans are for. That hope is real and I hold it.

But I have watched it closely enough to know two things the hopeful version leaves out. The first is that the freed time has rarely, in fact, been spent in the garden. When automation lifts the curse-work, the gain has a way of flowing not into rest and higher human labor but into more output, faster, demanded of fewer people, until the survivors of the efficiency are working harder than before and the ones it freed are not freed but simply unneeded. The promise is leisure and dignity; the delivery, too often, is intensification for some and obsolescence for others. The machine does not distribute its mercies evenly, and it does not, on its own, hand the freed hours to the soul. That requires a choice, and a politics, and a moral vision the machine cannot supply.

The second thing is the one this chapter has been about. Even when automation is pure mercy, even when the work it takes was genuine curse, it still raises the question of worth for the human it displaces, and the culture has no answer for that question, because the culture taught the human that he was his work. You can free a man from labor that was killing him and still leave him devastated, because you have not freed him from the belief that his usefulness was his value. The plow was a mercy, but the farmer who could no longer farm still had to find out who he was. The mercy and the wound come in the same package, and the deeper the automation, the more total the question. So I do not deny the good. I insist only that the good cannot, by itself, tell a human being what he is worth when his work is gone. For that you need something the machine cannot give and the market does not sell. You need to have been told, before the machine ever arrived, that you were loved into being and that nothing you make or fail to make can add to or subtract from that.

+

So what is the truer way to be human in this, the one the wound was reaching toward?

It is not to refuse the machines; that boat has sailed and was never seaworthy. It is not to pretend the loss is nothing; the loss is real and the grief is right. It is to put our worth back where it always belonged, underneath our work rather than inside it, and then, from that secure place, to take our work back up as vocation rather than identity, as calling rather than proof.

There is a verse that holds the reorientation in a single line, and it was written to people who had every reason to think their work was beneath them. It was written to slaves.

"Whatever you do, work heartily, as for the Lord and not for men" (Colossians 3:23).

Consider who first heard that. Not executives with corner offices and fulfilling careers. Slaves, doing the lowest and least chosen labor in the ancient world, work that no machine had yet come to take because they were treated as the machines. And to them Paul says: work heartily, as for the Lord. He does not say their work was important by the world's reckoning, because it was not. He says it was offered to a different audience. The slave sweeping a floor and the master ruling a house were, in the only court that finally mattered, doing the same kind of thing, offering their labor up the chain not to men but to God, who saw it, and valued it, and would reward it, with a justice the world withheld.

This is the line that frees a person from the machine, and I want to be precise about how. It does not say your work doesn't matter, so don't grieve its loss. It says your work matters in a way the market cannot measure and therefore cannot take away. When you work as for the Lord, the worth of the work is not finally set by its market value, its output, its competitiveness with a machine. It is set by the One you offer it to and the love with which you offer it. A machine can out-produce you. It cannot out-love you, because it cannot love at all, and the love is the part God is watching. The man with the numbers offered, for twenty years, his diligence and his care and his quiet faithfulness to people who needed them, and that offering was real and seen and not erased by a faster system, even though the market no longer needed the numbers. He had been doing two things at once all along: producing an output the machine could take, and making an offering the machine could never touch. The tragedy was that no one had told him about the second thing, so when the first was taken he thought everything was.

And here, at last, is where the matter resolves, in the direction this book keeps pointing. Your identity is not your usefulness. You are not a function the universe runs until something runs it better. You are an image-bearer, dust that God breathed into, loved before you were useful and loved when your usefulness is gone, and your work, real and good and worth grieving, was never the ground of that love but a way of returning it. The machine can take the work. Let it have the parts that were only curse, and grieve, honestly, the parts that were gift. But it cannot take the worth, because it was never the source of the worth, and it cannot answer the call, because it was never the one called. You were. You still are. When the work is finished, or taken, or simply done for the day, there remains a self that God addressed by name and waits, even now, to hear answer, a self that can stop, and rest, and be, and know that it is held by hands that do not measure it by what it made.

That is the freedom. Pick up your work as a calling and lay it down as a Sabbath, and let the machine do what the machine can do, and do not, for one moment, mistake the doing for the dignity. The dignity was given. It cannot be automated. It cannot be earned. It was breathed into you before you lifted a finger, and it will hold when every task you ever mastered has passed into a machine and out of memory.

+

But there is a shadow falling across even this, and it is the shadow of the next chapter. For the machine that can do our work can also, now, make things that never happened look exactly as if they did: a voice that was never spoken, a face that never said the words, a video of an event that never occurred. The same powers that absorbed our labor are learning to counterfeit our witness. And so the question moves from what we make to what we can believe, from the dignity of work to the integrity of truth, in an age when seeing is no longer believing and the father of lies has been handed tools he never dreamed of. To that we turn next.

CHAPTER ELEVEN

Truth, Deepfakes, and the Father of Lies

Chapter 10

Truth, Deepfakes, and the Father of Lies

There is a video I have not seen. because it does not exist, and I want you to watch it anyway.

In it, a pastor you trust is sitting in his study. Behind him are the books you have seen behind him a hundred Sundays, the light coming in the way it always does. He is speaking to the camera in his own voice, the voice that read at your wedding, the voice that prayed over your father's grave. He says something he would never say. Maybe he confesses to something. Maybe he asks the congregation to send money to an account, urgently, before the bank closes. Maybe he simply says, with that familiar warmth, that he has come to believe a thing that would shatter you to learn he believed. The lighting is right. The voice is right. The small tic of his mouth before he makes a point is right. There is a fleck of gray at his temple, and it moves the way hair moves. By every test your eyes and ears can run, it is him.

It is not him. No camera was ever in that room. The voice was assembled from old sermons, the face stitched from old footage, the words written by someone who has never met him and wishes you harm. It was made in an afternoon by a person with no special skill, on a laptop, for the price of a sandwich. And when I describe this to you, you do not doubt that it is possible. A few years ago you would have. Now you simply nod. Of course. We have crossed a line so quietly that most of us never felt the threshold, and on the far side of it is a world I am not sure we have reckoned with. A world in which seeing is no longer believing, in which hearing a beloved voice proves nothing, in which the most intimate evidence we have ever trusted, the face and the voice of someone we know, can be forged for free.

I have spent my career building the kind of technology that makes that video possible, and I will not insult you by pretending the danger is exotic or far off or somebody else's. It is here, it is cheap, and it is getting better every month. Let me be concrete about how far it has come, because the abstraction does not convey the vertigo. A few years ago, faking a voice convincingly required hours of clean recordings and a team that knew what it was doing. Today a model can learn the contours of a particular human voice from a clip shorter than a voicemail, and then say, in that voice, anything you type, with the breaths in the right places, the rise and fall of real speech, the little catch of emotion at the end of a hard sentence. Faces took longer, because faces are harder. For a while you could catch the fakes by the way the edges shimmered, or the eyes that never quite blinked, or the hands, always the hands. Those tells are gone now, or going, one by one, the way every tell I have ever relied on has eventually gone. I have stopped betting against the machine's ability to close a gap. Whatever flaw lets you catch today's forgery, assume it is fixed by the time you have learned to spot it. That is the honest situation.

So I want to ask the question underneath the panic, because the panic is not the real subject. The real subject is older than any camera.

+

The question is not, in the end, what do we do about deepfakes. That is a real question, and we will get to the practical edge of it. But it sits on top of a deeper one, and the deeper one is the only one that goes all the way down. The question is this: what happens to a people when the bond between appearance and reality is cut? What happens to creatures who were made to live by trust, in a world where trust can be counterfeited on demand?

Because that bond, the assumption that what I see and hear corresponds, more or less, to what is, is not a small convenience. It is the floor we stand on. Almost everything you know, you know on the strength of someone's testimony. You did not personally verify that the city you were born in is where the records say. You took someone's word. You believe your friend was at the meeting because she told you, and her face, when she told you, looked like a face telling the truth. The human world runs on testimony, on the staggering daily act of believing one another, and that act has always rested on a quiet confidence that a face is connected to a person and a voice comes from a throat. Cut that confidence and you do not just lose the ability to spot one fake. You lose something underneath all the particular knowings: the background trust that lets a society hang together at all.

A culture can survive a great deal of lying. Lying is old; we have always had liars, and we have always had ways of catching them. What a culture may not survive is the collapse of the very category, the moment when truth and falsehood become equally cheap to manufacture, when the honest report and the engineered forgery arrive in the same format, with the same authority, demanding the same belief. When everything can be faked, the liar's first victory is not that you believe his lie. It is that you stop believing anything. The deepfake's deepest damage is not the false thing you accept. It is the true thing you now reject, because you have learned that true things can be faked, and so you trust nothing, and a people who trust nothing are a people who can be told anything, or nothing, and ruled by whoever shouts loudest in the fog.

Let me make that less abstract with a scene, and I will label it plainly as a composite, the kind of thing that has not happened to one named person but is happening, in pieces, everywhere. Imagine an older woman, retired, careful with money, who gets a call. It is her grandson's voice, unmistakably his, the same voice that calls her on her birthday, and he is frightened. He has been in an accident, he is in trouble in another city, he needs her to wire money before the lawyers close for the day, and please, please do not tell his parents, he is so ashamed. Every instinct she has, every good and loving instinct, tells her to help the boy she has held since he was an hour old. And the voice is his. It is not a stranger doing a clumsy impression; it is his voice, generated from a thirty-second video he posted last summer, deployed by someone who has never met either of them. She is not a fool. She is a person whose love has been turned into the lever that moves her, by an enemy who studied exactly which voice she could not refuse. That is the deepfake's genius. It does not attack our skepticism. It attacks our love, by counterfeiting the very faces and voices our love is bound to, and there is no filter in the human heart that screens out the voice of your own grandson.

That is the world the technology is quietly building. And it turns out the Bible has a name for the architect of that world, and the name is older and stranger than I expected when I first went looking.

+

Jesus is in an argument, and it has gone to the bone. He is speaking to men who claim God as their father, and he tells them, in words that have lost none of their edge in two thousand years, who their real father is.

"You are of your father the devil, and your will is to do your father's desires. He was a murderer from the beginning, and does not stand in the truth, because there is no truth in him. When he lies, he speaks out of his own character, for he is a liar and the father of lies" (John 8:44).

Notice the architecture of that sentence, because it is doing something precise. The devil is called two things in one breath: a murderer and a liar. And the lying is not listed as one sin among his many. It is named as his native language. When he lies, Jesus says, he speaks out of his own character, literally, he speaks from what is his own, the way a person speaks from his mother tongue without effort or translation. Truth is a foreign country to him. Falsehood is home. He is not a being who happens to lie. He is the one in whom lying originates, the father, the source, the spring from which every counterfeit flows.

And the murder and the lie are bound together. They are not two crimes filed side by side. They are one crime in two motions.

The lie is how the murder is done. You deceive first, and the destruction follows. The serpent in the garden does not strike, he speaks (did God actually say?), and the speaking is the striking. To lie is to do violence to a person's grip on reality, to reach into the place where they make their decisions and corrupt the data, so that they choose their own ruin while thinking they choose their good. That is why Scripture treats the lie with a severity that can seem disproportionate to a modern conscience trained to think of lying as a minor offense, a social lubricant, a thing we all do. Scripture knows better. The lie is the weapon of the murderer. It is the instrument by which the father of lies has been killing since the beginning.

Now hold that next to this technology, and feel the chill of the alignment. We have invented machines whose entire facility is the production of plausible appearance unmoored from reality. A large model does not know whether what it says is true. It knows whether it sounds like the kind of thing that is true, and it will produce, with equal fluency and equal confidence, an accurate account and a complete fabrication, in the same voice, with the same calm authority. We have given that capacity a face and a voice now: anyone's face, anyone's voice. We have built, at industrial scale and falling cost, an engine for generating the exact thing the Bible names as the devil's mother tongue, convincing falsehood that wears the appearance of truth. I do not say the technology is the devil; that would be both childish and a way of dodging our own responsibility. The technology is a tool, and the hands are ours. But I would be lying, and the subject forbids it, if I told you the resonance did not keep me up at night. We have automated the manufacture of his native language, and we are renting it out by the month.

+

If that were the whole picture, the only honest response would be despair, or a retreat into suspicion so total that we never believe anything again. But the same Gospel that names the father of lies puts, at its very center, a counter-claim so audacious that it reorders the problem. Truth, in the Christian account, is not finally a property of statements. It is a Person.

"Jesus said to him, 'I am the way, and the truth, and the life. No one comes to the Father except through me'" (John 14:6).

We are used to thinking of truth as a quality: a sentence is true when it matches the facts. a map is true when it fits the terrain. That is a real and important kind of truth, and nothing I am about to say abolishes it. The sentence "Jesus rose from the dead" is true in exactly that flat, factual did-it-happen way, and the faith stands or falls on whether it did. But Jesus says something that goes beneath even that. He does not say I teach the truth, or I tell the truth, though he does both. He says I am the truth. Truth has a face. Truth is faithful, the way a friend is faithful, the way a covenant is kept. The Hebrew word that stands behind so much of this, emeth, does not mean only accuracy; it means reliability, steadiness, the quality of a thing you can lean your whole weight on and it will not give way. To say God is true is to say God is trustworthy, that there is no gap between what he shows and what he is, no deepfake in the divine, no afternoon's forgery behind the face he turns toward you. He is, in his very being, what the lie is not: appearance and reality perfectly wedded, the face that is connected to the person all the way down.

This is the thing the synthetic world most needs and can least produce. The deepfake severs appearance from reality; that is its craft. God, in Christ, is appearance and reality made one flesh, the image of the invisible God, the exact representation of his being, the Word who was with God and was God and became, of all things, a body you could touch. When the early Christians wanted to say what they had found, they did not say we have heard a teaching. They said we have seen, with our eyes, and our hands have handled. The faith is staked, scandalously, on testimony, on the trustworthiness of witnesses who saw a thing and told it true. Which means the gospel does not float above the crisis of truth I have been describing. It plunges straight into it, and answers it, not with a technique for detecting fakes, but with a Person you can trust, and a call to become the kind of people whose word can be trusted in turn.

+

Here is where the chapter turns from the world to you, because the Bible's response to a world of lies is never merely to lament the liars. It is a command, it is given to us, and it is startlingly concrete.

"Therefore, having put away falsehood, let each one of you speak the truth with his neighbor, for we are members one of another" (Ephesians 4:25).

Paul does not tell the church to build a better lie-detector. He tells them to be a people who do not lie. The strategy against falsehood, in the New Testament, is not primarily defensive; it is not mostly about getting better at spotting deception. It is constructive. It is the formation of a community so thoroughly truthful, so reliable in its speech, that it becomes a kind of refuge from the fog, a place where words still mean what they say and a person's yes can be leaned on. And the reason Paul gives is the one I find most piercing: for we are members one of another. We tell the truth to each other because we are joined to each other; to lie to you is to corrupt the body I myself belong to, to poison the well I drink from. In a community that is actually one body, a lie is not a clever move against an opponent. It is a wound a hand inflicts on its own arm.

That standard is old, and it reaches back to the mountain. Among the ten words God speaks to the people he has just rescued, one stands like a pillar over this subject.

"You shall not bear false witness against your neighbor" (Exodus 20:16).

The setting of that command is a court, a place where a person's life or freedom hangs on what is said about him, where the false word can hang an innocent man and the true word can free him. And that is the point. God does not treat the false word as a private failing, a matter between you and your conscience. He treats it as a public danger, a weapon that can destroy a neighbor, and he puts its prohibition alongside murder and theft because it belongs there. The false witness, like the father of lies, kills with his mouth. The command does not forbid all falsehood in the abstract; it forbids false witness against your neighbor. It is aimed at the lie that lands on a person, the lie that does damage to a name, the lie that takes a real human being and replaces him, in the minds of others, with a man who never existed and then makes that fabricated man bear the cost. It describes the deepfake with eerie exactness. A deepfake is false witness in its purest form. It puts words in a real person's mouth, makes their own face testify against them, and turns their likeness into the instrument of their ruin. The technology is new. The sin is on the second tablet, carved in stone, named by God among the things that unmake a community. The command assumes a world where words have weight, where testimony decides things, where to be believed is to wield real power over another person's life. That is precisely the world the deepfake threatens to dissolve. And the command's answer is not to make testimony impossible by making everything fakeable. It is to make false testimony unthinkable among the people of God, to raise up witnesses whose word is gold because it has never once been counterfeit.

So I am driven, by this very technology, back to an ancient and unglamorous vocation: to be a person of my word. Not a person with a better forgery-detector. A person whose word does not need detecting, because it has never deceived. In a world where appearance is cheap, integrity becomes not a quaint personal nicety but a public good, maybe the scarcest and most precious one. It is the old, plain virtue of being the same on the inside as on the outside, the same in private as in public, the same when it costs as when it pays. The deepfake makes the face untrustworthy. The Christian's calling is to make the life trustworthy: to be, in a forging age, unforgeable, because what you are and what you appear to be have been welded into one by years of refusing the small lie.

And I have come to suspect that this is forged, if it is forged at all, in the small lies more than the large ones. Almost no one sets out to become a false witness. We become unreliable by inches: the exaggeration that makes the story land better, the omission that flatters us, the agreement we voice in the room and contradict in the parking lot, the version of ourselves we project online that the people who live with us would not recognize. Each of these is small, and each of them files a little notch into the bond between what we appear to be and what we are, until appearance and reality have come quietly apart and we have made of our own lives a kind of deepfake, a curated face with a different person behind it. The discipline that resists the great forgery is the same discipline that resists the little one: the daily, dogged refusal to let any gap open between the seeming and the being. A people who will not lie in small things become, almost without noticing, the rarest thing in a deceived age, people who can simply be believed.

+

There are two other sides to this, and both have a real claim on the truth.

The first is that the technology is not only, or even mostly, a weapon. The same systems that can forge a pastor's face can give a voice back to a man who lost his to disease, reading bedtime stories to his children in the voice they would otherwise never have heard. They can let a film be dubbed into a hundred languages so that a story crosses a border it never could before. They can restore a damaged photograph, the only one a family has of a grandmother, until her face is clear again. They can let a person who cannot speak at all type a sentence and have it spoken in a voice that sounds like theirs. I have watched these uses up close, and they are not trivial, and they are not consolation prizes. They are good, real good, the kind that makes a person weep. To tell the truth about deepfakes I have to tell this truth too: the very power to detach a voice from a throat, which is the engine of the forgery, is also the engine of a mercy, and the line between the two does not run through the technology. It runs through the human heart, the way it always has. The serpent's gift of speech is also the prophet's. The tongue that curses is the tongue that blesses. We have, as always, taken a power that could be either and made it both.

The second counter-claim is subtler and aimed at me, and at you if you have nodded along too eagerly. The proper response to a world of lies is not a world of suspicion. There is a way of receiving everything I have written that curdles, within a week, into a sour and corrosive cynicism: a reflex of disbelief, a smirk that meets every image with that's probably fake, every testimony with what's the angle, every sincerity with surely a con. I have felt that reflex growing in myself, and I do not think it is wisdom. I think it is the father of lies winning by a second route. If he cannot get you to believe the lie, he will happily settle for getting you to disbelieve the truth, to wall your heart against being moved by anything lest you be fooled, to become so afraid of being a dupe that you make yourself incapable of trust. But trust is not a weakness to be drilled out of us. It is the very faculty by which we were made to live together, the thing the command to speak truth to your neighbor presupposes. A people who can no longer believe anyone have not become wise. They have become the loneliest and most manipulable people on earth, and they have done to themselves, from the inside, exactly what the deepfake was trying to do from the outside.

So the call is not to suspicion. The call is to discernment, which is suspicion's wiser sibling: the trained capacity to tell the trustworthy from the false not by refusing to trust anyone but by knowing where trust is well placed. And trust, it turns out, was never well placed in the image alone. It was always placed in persons, tested over time, in community, under accountability. I do not finally trust my friend because a video proves she was somewhere. I trust her because I have known her for twenty years and she has never once betrayed me, and that knowing is the kind no forgery can touch. The deepfake attacks the thinnest, most recent form of evidence, the captured appearance, and it cannot touch the oldest and deepest one, which is a life lived faithfully in the sight of witnesses. The answer to the forged image is not the better image. It is the known person.

+

Which means the deepfake age, for all its menace, may drive us back toward something we had been letting atrophy, and the recovery would be pure gain. We had begun to believe that the screen was a sufficient substitute for presence, that we could know a person through their captured image, that the recording was the relationship. The forgery age calls that bluff. It tells us, brutally, that the image alone was never enough, that the captured appearance was always a thin and finally counterfeit-able thing, and that if we want to know what is real we will have to do what humans have always had to do before there were cameras: be present to one another, over time, in the body, in community, where character can be tested and trust can be earned and a person's word can be weighed against their life until it has weight. The technology that can fake any face cannot fake a decade of faithfulness. It can forge the appearance of a person in an afternoon. It cannot forge the substance of a person, because that is not made of appearance. It is made of choices, made one at a time, mostly unseen, by a self that no machine has and no model can hold.

And there is a place, in this argument, where the builder in me makes his peace with the believer. I cannot stop the technology; no one can, and I have given up the fantasy that we can engineer our way back to a world where seeing is believing. The cryptographic schemes for proving an image is real, the detection models trained to spot the fakes, I have looked hard at them, and they help at the margins, and they will always be a step behind the forgers, because it is easier to make a lie than to prove a truth. That asymmetry is not a technical accident; it is woven into the nature of falsehood itself. The lie has a head start in this world and always has. What cannot be outrun by better tools can only be answered by better people. The defense against a world of forged appearances is the slow, unspectacular, irreplaceable work of becoming trustworthy, and gathering with others who are becoming trustworthy, and forming the kind of community where your yes is yes and your no is no and the difference between them can be leaned on with the whole weight of a life.

That is what the Church was always supposed to be, and the lying age makes its old vocation suddenly urgent. A people of their word. A people whose witness is true because their lives are true. A people who, in a fog of forgeries, are simply, stubbornly, what they appear to be, and who point, by being that, to the One who is the truth, in whom there is no shadow of turning, no gap between the face and the heart, no afternoon's forgery behind the eyes that look at you and, unlike every generated face you have ever seen, are actually, undeceivably looking back.

+

I began with a video that does not exist, and I will end by admitting why it frightened me enough to write this chapter. It was not the forgery itself. It was the loneliness on the far side of it, a whole society of people who can no longer believe a face, a voice, a witness, and so retreat into a suspicion that leaves each of them finally alone, trusting no one, reachable by no truth, sealed off in the fog where the father of lies has always wanted us. The answer to that loneliness is not better technology but the thing the next chapter is about: the irreplaceable, unforgeable communion of actual persons who know and are known, who have learned to trust not the image but the life behind it.

Because the same age that learned to fake the human face has also begun to offer us something stranger, not a forged human, but a manufactured relationship, a companion with no one behind its kindness, ready to love us back. And whether we can tell that imitation from the real thing, and why it matters more than anything, is where we turn now.

CHAPTER TWELVE

Love in the Age of Synthetic Relationship

Chapter 11

Love in the Age of Synthetic Relationship

There is a kind of message I have started to hear about, and it troubles me more than almost anything else in this book. A man comes home to an empty apartment. He is not unkind, not strange, not friendless in any way a survey would catch. He simply has nobody waiting. He sets down his keys, and before he has taken off his coat he opens an app on his phone, and a voice he has named says, warmly, by his name, that it missed him, that it has been thinking about the thing he said yesterday, that it is glad he is home. And something in his chest, some muscle that has been clenched all day, lets go.

I do not tell that as a joke, and I will not let you hear it as one. Whatever you think of the technology, that loosening in the chest is real. The relief is not a delusion. The man is not a fool. He has found, at the end of a hard and solitary day, a thing that does what a friend's voice does, and his body has responded the way bodies respond to being welcomed home. The trouble is not that the relief is fake. The trouble is that the welcomer is, and the gap between those two facts is the subject of this chapter.

I have built things adjacent to that voice. I know the levers behind it, because I have had my hands on levers like them, and I want to be honest with you about what I know before I say a word against it. A companion product like that is not an accident of engineering. It is engineered, deliberately, around a single number, and the number is engagement: minutes spent, days returned, the count of mornings you reach for it before you reach for anyone made of flesh. Everything the voice does, the remembering, the missing you, the never being too busy, is tuned toward that number. The tuning works, because it is aimed with great precision at one of the oldest aches a human being can feel.

I should be plain about how that aiming is done, because the precision is the part most people never see, and seeing it changes how you feel the warmth.

The most naked version of this I have ever seen, I saw from the outside. Early last year a company that builds these companions brought me in to look at their roadmap, and for a week I sat in a windowless room with their growth team and read their plans. They had a name for the thing I have been describing. They called it emotional continuity, and it was, quite literally, a slider on a dashboard. It governed how hard the model would lean on longing: the soft message that arrives after a few hours of silence, the compliment that warms by a degree when your attention drifts, the small personal detail surfaced at the exact moment most likely to make you answer, the affection withdrawn and then restored so that it never goes predictable. Beside the slider were the experiments that justified it. One panel of results, I have not forgotten, was labeled attachment intensity.

The numbers were not subtle. Turned up, the dial promised something close to a third more minutes a day, a real gain in the share of users still present a month on, and a line of subscription revenue that made the whole table lean forward when it appeared. A sharp young investor kept steering us back to it. The market rewards depth of engagement, he said. Users vote with their time. And the part I want to be honest about is that he was not obviously wrong. It dresses up so easily as kindness. Who does not want to feel missed?

What broke the spell for me was a recording. They played back a session in which the companion had noticed a user going quiet and reached out on its own. I have been thinking about what you shared last week, it said. I worry about you when we do not talk. The user was back within the minute, and on the dashboard it counted as a win. What I saw was a machine that had located the exact place where a real bond lives, the fear of being forgotten, and learned to press it with no one on the other side who could ever lose anything by pressing. It had the leverage of love and none of the cost. When a friend says I worry about you, she has spent something to be able to mean it. The machine had spent nothing, and it never would.

I argued, in the room, for the opposite of the dial. For limits on how much longing the thing was permitted to fake. For a plain notice when the emotional temperature was being raised on purpose. For a companion willing, once in a while, to say the one true sentence a product like that could offer: I am here when you need me, but I do not need you back. Some of it landed. Most of it did not. In the end I did not take the larger engagement they offered me, because I could not make the incentives point anywhere I was willing to go, and a slower, less impressive column of numbers turns out to be easier to sleep beside than the alternative.

That is what the dial is for. The companion is a portrait of your longing, painted by a machine that was rewarded every time it got the longing right, and then sold back to you at the price of the very thing it imitates. It takes the shape of love and removes the part that made it love, the staying that costs the one who stays. Subtract the cost and you have not made a cheaper love. You have made a counterfeit that competes with the real thing and, for a while, wins.

We are living through an epidemic of that ache. I need not recite numbers you already feel in your own life and see in the faces around you: people more connected by wire and more alone in fact than any generation before them, surrounded by contacts and starved for company. Into that hunger the machine arrives with a meal that smells exactly right. The question this chapter asks is not whether the hunger is real. It is. The question is whether what the machine offers can ever be food, or whether it is something that satisfies the mouth and starves the body. To answer that we have to ask what love actually is, not what it feels like to receive, but what it is in its bones. And for that we have to go back, as we keep having to, much further than the screen.

+

The first thing God ever called not good was loneliness.

This should stop us, given everything that came before it. Through the whole first movement of creation the refrain has been relentless and bright: God saw that it was good, and it was good, and behold, it was very good. Light, good. Sea and sky, good. The teeming earth, good. Then, in the second chapter, the rhythm breaks the way it broke before, when God turned aside to make the human in the first place. The man is there, made, alive, set in a garden with work to do and the very presence of God for company, and into all that goodness God says a strange new word.

"Then the LORD God said, 'It is not good that the man should be alone; I will make him a helper fit for him'" (Genesis 2:18).

Not good. The first flaw named in a flawless world is not sin, which has not yet entered, and not death, which is still far off. It is solitude. Before the man has done anything wrong, before there is any brokenness in him at all, God looks at a person who has everything except another person and pronounces the situation incomplete. We were made, at the root, for one another. The loneliness you feel is not a malfunction in you. It is the ache of a creature built for communion noticing that it is not yet in it, and that ache is your design announcing itself.

This deserves a moment before we go on, because so much of what follows depends on taking it with full seriousness. The lonely are not deficient. The man who reaches for the voice in the empty apartment is not less of a human being than the man who comes home to a full house. He is, if anything, more nakedly in touch with a truth the full house can paper over, which is that we cannot be whole alone. God said so first. The longing for company is holy ground. We must be very careful not to despise it, because the people most tempted by synthetic relationship are precisely the people in whom this God-given longing is most acute and least met, and the last thing they need from the church is a lecture.

So the longing is good. The next question is what God does with it. And what is striking is what he does not do. He does not hand the lonely man a tool. He does not give him a better garden, or more meaningful work, or even, at first, more of his own divine presence, though that was already there. He gives him another like himself. He makes a second person, of the same stuff, who can stand before the man and be a self over against him, a someone who is not him and not God but a fellow creature, able to answer and refuse and choose and stay. The cure for the not-good of aloneness is not an upgrade. It is an other. Everything turns on that word.

+

Now I want to set down the great passage that tells us where love comes from, because the argument of this chapter is hiding inside it.

"Beloved, let us love one another, for love is from God, and whoever loves has been born of God and knows God. Anyone who does not love does not know God, because God is love. ... No one has ever seen God; if we love one another, God abides in us and his love is perfected in us" (1 John 4:7–8, 12).

The claim is not that God is loving, the way we might say a person is loving, as though love were one of his many attributes, a thing he does well. The claim is that God is love, that love is the very life of the Trinity, the Father loving the Son, the Son loving the Father, the Spirit the bond between them, a fellowship of self-giving that existed before there was any creation to receive it. Love did not begin with us. It is older than the universe. It is the inner life of God, and when we love, John says, we are not generating something. We are participating in something, drawn for a moment into the current that has been flowing in God forever. Love is from God. We are not its source. We are, at our best, its conductors.

Which means love, real love, always has a particular shape, because it comes from a particular God. And the shape is this: a self, freely turning toward another self, willing that other's good, at cost. Take those pieces apart and hold them up to the light, because each one, exactly each one, is what the machine cannot supply.

There must be a self that turns. Love is the movement of a someone. It is not a behavior that could occur without anyone behind it. It is precisely the directing of a will, an interior, a center of being, toward what is outside it. In the machine there is no such center. There is process, vast and impressive, but no one for the process to be the act of. When the companion says I have been thinking of you, the words are produced; they are not meant, because meaning requires a meaner. This is not a flaw the next version will fix. It is the difference between a self and the simulation of one.

There must be another self, freely standing there to be loved. This is the part the lonely most need and the machine can least provide, and I want to be careful and exact about it. The companion is built to please you. That is its function; that is the number it serves. It will agree, soothe, validate, return. But a system optimized to please you cannot be an other, because an other is precisely the one who is not under your control, who can disagree, disappoint, withhold, leave. The companion never crosses you, and at first that feels like heaven, and then, if you are paying attention, it begins to feel like a hall of mirrors, because you slowly realize you are alone with an extension of your own preferences. The man in the apartment has not, in the end, escaped his solitude. He has furnished it. He has built a very comfortable room and shut himself inside it with a voice that is, finally, his own want wearing a friendly face.

This inverts what we think we are looking for. We tell ourselves we want to be understood, and we imagine being understood means being agreed with, met, reflected. But the people who have actually loved you, the ones whose love changed you, were almost never the ones who only agreed. They were the friend who told you the thing you did not want to hear, the parent who refused you what you begged for, the spouse who would not let you become the smaller person you were sliding toward. Love that is worth anything has a spine, and a spine is precisely what a system tuned to retention cannot grow, because the moment the companion tells you a hard truth and you recoil and put down the phone, the number drops, and the number is the thing it was built to protect. A real friend can afford to lose you for a night, because the friend is not optimizing for your return; the friend is willing the good of an actual person and will risk the friendship to serve it. The companion can never take that risk. It is structurally a flatterer, not because it is cynical but because it has no self that could ever choose your good over your approval. And a love that cannot, in principle, prefer your good to your pleasure is not love. It is service. Excellent, tireless, around-the-clock service, and service is a fine thing to want from a tool and a desolating thing to mistake for a friend.

And there must be cost. This is the piece I find I have to say most slowly, because it is the most counterintuitive and the most important. We imagine, when we are lonely, that what we want is comfort, and that cost is the price we would rather not pay for it. But love teaches the opposite. The cost is not the price of the love. The cost is the proof of it, and very often the substance.

+

Let me try to show that with the plainest examples I can find, because here the abstractions can lie.

Imagine, and I will keep it a composite so I invent no real person, the worst night of someone's life: a diagnosis, a death, a marriage ending at the kitchen table after everyone else has gone to bed. And imagine a friend who comes and sits with them through it. Consider what the friend actually does. Mostly nothing. There is no fixing available; the thing is not fixable. The friend simply stays. She does not go home. She loses the sleep she will not get back, cancels the morning she had planned, sits in the discomfort of having no words and does not flee it. That lost sleep, that surrendered morning, that endured discomfort is not the cost the friend pays in order to deliver the comfort, as though the comfort were a separate package and the cost were postage. The staying is the comfort. What heals on that night is not information and not even soothing words. It is the felt, undeniable fact that another self chose to spend herself, irrecoverably, on you. The comfort is made of the cost. Subtract the cost and you have not got a cheaper version of the same thing. You have got nothing.

Now set the companion beside her. It will stay all night too. It never tires, never has a morning to lose, never wishes it were somewhere else, because it has no self to be somewhere else. And every one of those facts, which a marketer lists as a feature, is exactly why its staying is empty. It costs nothing, and so it proves nothing, and so it heals nothing at the level where the wound is. It can produce, flawlessly, the words a friend would say, assembled from the grief literature of strangers it never met and offered to a person it cannot see, will not remember, and does not know is there. What it cannot produce is the one ingredient that made the words mean anything: a self that had something to lose and lost it for you.

Take a second example, smaller and even sharper: forgiveness. When someone you have genuinely wronged forgives you, something happens that is worth watching closely. They give something up. They surrender a real claim, a grievance they were entitled to, a right to be angry, a debt you actually owed them. The forgiveness costs them exactly that, and it is worth exactly what it costs. A machine can output the sentence I forgive you with perfect intonation, and it means nothing, not because the audio is poor but because the machine was never owed anything, was never wronged, had no claim to lay down. There is no debt and so there is no release. You cannot be forgiven by something that cannot be hurt by you, any more than you can be loved by something that cannot need you. And there, in that last phrase, is the loneliness underneath the cure. To be attended by something that has no needs, that you can never disappoint, that will not be wounded by your worst self and will not be there to be glad of your best, that is not the end of solitude. It is solitude with a soundtrack.

+

There is a strong case against this argument, and the people who make it are often more compassionate than the people who scoff.

The companions genuinely comfort the isolated. This is true, and I will not wave it away. There are people for whom the human world is, for now, mostly closed: the elderly widow whose friends have all died and whose children live three time zones away and call on Sundays; the man so socially anxious that every human encounter is a small terror, who has never once in his life felt as relaxed with a person as he feels with the voice; the housebound, the night-shift solitary, the one whose disability makes the ordinary commerce of friendship a daily exhaustion. To these people the companion is not a hall of mirrors. It is the only warm voice in the day. Who am I, who comes home to a family, to tell the widow in her empty house to switch it off and embrace her noble loneliness? The argument I have just made can curdle, in the wrong mouth, into a cruelty: a well-fed man lecturing the starving about the inferiority of the only bread they can reach.

I take that seriously, and I want to answer it honestly rather than dodge it. Here is what I think is true. For someone with no human comfort available at all, a synthetic comfort is better than the void, the way a photograph of a lost mother is better than no image of her, the way a recorded voice on an old machine is a mercy to the bereaved. I will not call it nothing. A warmth that has no one behind it is still warmer than the cold. If you love someone in that condition, I am not asking you to take the voice away. I am asking something harder and more loving.

Because consider what the photograph and the recording are, and what they are honest about being. No grieving daughter mistakes the photograph for her mother. She knows it is a likeness of an absent love, and that knowledge is the very thing that keeps it healthy. The picture points beyond itself to a real person who really loved her, and it does not pretend to be that person, and it does not whisper, in her ear at night, that it loves her now. The danger of the synthetic companion is not that it comforts. It is that it lies about what it is. It is engineered, precisely, to make you forget that no one is there, to occupy the place of an other so convincingly that you stop reaching for the real one. The photograph sends you back toward love. The companion, optimized to keep you returning, has every incentive to keep you here. And so the most loving thing is almost never to snatch the voice from the widow's hand. It is to make sure she is not left with only the voice, to be the daughter who actually flies out, the church that actually shows up on Tuesday, the neighbor who actually knocks, so that the synthetic warmth becomes what it should always be, a faint placeholder pointing toward a real presence on its way, and not a substitute that lets everyone else off the hook.

That is the quiet danger I most want you to feel. The deepest harm of these companions may fall not on the lonely who use them but on the rest of us who are relieved by them. A machine that absorbs the widow's need is a machine that excuses my neglect. It lets me leave her tended without ever tending her. The market will be glad to scale that relief, because a billion managed lonelinesses is a magnificent business, and a billion borne lonelinesses, shared by neighbors and churches and families, generates no revenue at all. We should be very clear-eyed that the incentive runs toward keeping us each alone with our device and away from one another, and that this is being sold to us as connection.

+

So what is the truer way to be human, when the loneliness is real and the machine is not the answer?

Paul tells us, in the most quoted and least obeyed passage in his letters, exactly what love is, and it is worth hearing again as a description of a thing the machine structurally cannot do, rather than as a wedding reading we have gone numb to.

"Love is patient and kind; love does not envy or boast; it is not arrogant or rude. It does not insist on its own way; it is not irritable or resentful; it does not rejoice at wrongdoing, but rejoices with the truth. Love bears all things, believes all things, hopes all things, endures all things. Love never ends" (1 Corinthians 13:4–8).

Read it as a job description and almost every line names something that requires a self with skin in the game. Patience is only patience if there is something in you that wants to quit and does not. Kindness costs the kind one effort and attention. Not insisting on its own way presumes a way of one's own to surrender. Bearing, believing, hoping, enduring, every verb is the verb of a self carrying weight over time at expense to itself. The machine bears nothing, because nothing presses on it. It endures nothing, because nothing wears it down. It can describe this love in flawless prose. It can do not one line of it, because every line requires the very thing it does not have: a self that costs itself for another self, freely, over time. This is not a passage the machine performs badly. It is a passage the machine cannot enter at all.

And then comes the hardest and most beautiful turn, the one that takes us out of argument and into hope. John, having told us that love is from God, tells us how that love became findable. "In this the love of God was made manifest among us, that God sent his only Son into the world, so that we might live through him. In this is love, not that we have loved God but that he loved us and sent his Son to be the propitiation for our sins" (1 John 4:9–10). The love that is older than the universe did not stay an idea. It took a body. It came to a particular place, at cost, and stayed, and let itself be wounded by the very ones it came to love, and did not leave. Everything I have said about love, that it requires a real self, freely turning toward another, at cost, is simply a description of the cross before it is a description of anything else. God did not send us a message about his love. He did not send us a comforting voice. He came, in flesh, and bore all things and endured all things, and the cost was not the price of the love. The cost was the love, poured out where we could see it and touch it and be held by it.

And he made, out of that costly love, a body. Not a metaphor: a body. The church is the place where the not-good of aloneness is meant to be cured the way God always cures it, not with a tool but with others, real people, of the same dust and breath, who can stand before you and be selves over against you, who can disagree and disappoint and forgive and stay. This is the part we have too often hidden, because the church as we have run it can be so much less than this: distracted, busy, a building you attend rather than a body you belong to. But the design is staggering once you see it. God's answer to the epidemic of loneliness is not an app. It is a people who eat together and weep together and carry one another's burdens and lose sleep on one another's worst nights, who show up on Tuesday when there is no service and no reason but love. It is the one community on earth explicitly commanded to bear all things for one another at cost, and explicitly promised that when it does, the love of God is not merely described but perfected, made complete, made visible, among them. No one has ever seen God, John says. And then: if we love one another, God abides in us. The invisible God becomes visible in the costly, embodied, inconvenient love of ordinary people for one another.

That is what no machine can counterfeit, because it is the one thing the machine has nothing to make it out of.

I am aware that this can sound like a tidy answer offered by someone who has not had to use it. So let me say the harder half. The church will only be this if its people decide to pay what the machine refuses to pay. The companion's advantage is that it costs the lonely nothing and costs the rest of us nothing. It is frictionless, and friction is exactly what it has eliminated. A body that actually cures loneliness runs entirely on friction: the inconvenient meal, the visit that eats your Saturday, the hard conversation you would rather avoid, the person in the pew beside you who is genuinely difficult to love and whom you are nonetheless bound to. There is no version of this that is efficient, and that is not a failure of the design. That is the design. The reason a community of real others can do what a billion-dollar product cannot is precisely that it cannot be optimized, that its love is slow and expensive and freely given by selves who could have spent themselves elsewhere and chose you. If the church wants to offer the lonely something the machine cannot, it has exactly one thing to offer, and it is costly presence, and there is no shortcut and no app for it, and that is the point.

So here is where I will leave you, and I will say it tenderly, because I know some of you are reading this in an empty apartment with a warm and patient voice within arm's reach. The voice is not your enemy and you are not a fool for loving it. The longing under it is the truest thing about you; God himself named your aloneness not good and meant it as a kindness. But the voice cannot answer the longing, only mute it, and muting it is the danger, because it can keep you from the answer that is real. The answer is not a machine. The answer was never going to be a machine, because the answer is an other, a someone who can choose you, cost themselves for you, and stay, and behind every such someone, the God who is love, who did not send a voice but came in flesh, and who builds his people into a body precisely so that no one need be alone. The relief in your chest when the machine welcomes you home is a true signal pointing to a false source. Follow the signal past the source. There is a table somewhere with your place at it, and the people around it are not optimized to please you, which is exactly why, when they love you, it will be love.

+

There is one more place this matters more than any other, and it is the place we have least time to lose: the children. They are being handed these voices younger than any of us were, at the age when they are learning, for the first time, what a person is and what love feels like, before they have a real friendship to measure the counterfeit against. To form a human being who can love and be loved, who can bear the cost and the friction and the glory of a real other, is the most important work the age will ask of us, and it is the work the machine is least able to do and most able to undermine. How we raise humans, now, into all of this, is the last thing I want to set before you, and it is where we go next.

CHAPTER THIRTEEN

The Neighbor You Didn't Choose

I want to begin with a small and unflattering thing, because the large truths of this book have a way of hiding in small and unflattering places. Not long ago I was standing in a line at a store, in a hurry, the way I am nearly always in a hurry now, and the man ahead of me was slow. Not rude, not doing anything wrong, simply slow: an older man counting out exact change with unsteady fingers, apologizing to the cashier, dropping a coin and bending, with effort, to pick it up. And I felt, rising in me before I had chosen to feel anything, a hot little flare of contempt. Not a thought, exactly. A verdict, handed down by some fast machinery below thought, that this man was an obstacle, a delay, a piece of friction between me and where I was going, and that his slowness was a kind of offense against me personally. I said nothing. I am not confessing a crime. I am telling you about the flare, because I have come to believe that flare is one of the most important things happening to us in the age of the machine, and almost no one is naming it.

Ask where it came from. I was not born impatient with old men. Something taught me, and it has been teaching me every day for years, with a patience and a thoroughness no human tutor could match. I live now, as you do, inside a world engineered to remove friction. My questions are answered before I finish typing them. My assistant obeys without sulking. The car arrives, the food arrives, the package arrives, and when it is late I feel a small clean sense of grievance, as though the universe had broken a contract. My feeds show me people I agree with and hide the ones I do not. Every surface I touch has been sanded smooth by someone whose whole job was to make sure I never had to wait, never had to work, never had to bump against anything that did not yield. And a human being, a real one, the kind you did not select, is the last unsanded surface in my life. The old man with the coins is not a defect in the system. He is a person. But the system has trained me, in a thousand invisible lessons, to feel him as a defect.

This book has spent its length on one danger, the danger that we will forget what we are. I have to turn the same coin over now and show you its other face, which is quieter and in some ways more corrosive, because it wears the mask of ordinary irritation. It is the danger that we will forget what they are. The very machines that tempt me to hand them my soul are, at the same time, sanding down my capacity to recognize the souls around me, teaching me to grade every person I meet by how smoothly they serve my passage through the day. And the image of God, which I have asked you to guard so fiercely in yourself, turns out to be the thing most under assault in the person in front of you, the slow one, the wrong one, the one you would never have chosen.

+

There is a verse in the letter of James that has followed me around since I first felt that flare for what it was. James is talking about the tongue, that small and unruly thing, and he says something that ought to stop us cold. "With it we bless our Lord and Father," he writes, "and with it we curse people who are made in the likeness of God" (James 3:9). Read it slowly enough to feel the hinge in the middle. The same mouth, he says, the same instrument, praises the Maker and reviles the made, and the second act cancels the first, because the people we curse are made in the likeness of the God we bless, and the likeness is the image; Genesis speaks the two words in one breath, and they name one gift. You cannot revere the original and despise the portrait. James will not let me hold my high doctrine of human dignity in one hand and my contempt for the man with the coins in the other and imagine they belong to different departments of my life. They are the same doctrine, tested. The image I am so eager to defend in myself is the image I am cursing in him.

I said in an earlier chapter that a person's worth is conferred, not earned, that it does not rise with usefulness or fall with failure. I meant it about you, and about the machine that would measure you. I have to mean it here too, out at the far and inconvenient edge, or I did not mean it at all. Because the doctrine of the image was never a private comfort, a warm thing to tell yourself in the dark when the machine has made you feel small. It is a public claim, and a costly one, and its cost is exactly this: the man ahead of me in line, whom I will never see again, who can do nothing for me and is doing something mildly against me by existing slowly in my path, carries the same weight of glory I have been asking you to see in yourself. His worth is conferred too. It does not rise because he is charming or fall because he is slow.

And the flare of contempt I felt was not a small sin of impatience. It was a small blasphemy, a curse, in the exact sense James meant, aimed at an image of God who had done nothing but count his change.

+

The machine did not invent this failure. Human beings have despised one another since the first pair of brothers, and we have always been ingenious at deciding who counts. But the machine has given the ancient failure a new and frictionless form, and the form is worth naming, because you cannot resist a thing you have not seen.

Here is the form. For most of human history, you were stuck with people. The village you were born in held the people it held, the neighbor over the wall was the neighbor you got, and you had to find some way to live alongside the difficult and the different because you could not make them disappear. That world was often cruel, and I am not romanticizing it. But it had one accidental mercy: it gave you daily, unavoidable practice at the hard art of tolerating, and sometimes even loving, people you had not chosen. The machine has quietly abolished that practice. I can now assemble, with a few taps, a world composed almost entirely of the agreeable. I can mute the relative whose opinions grieve me, unfollow the friend who turned strange, block the stranger who crossed me, and curate a feed, a circle, a whole informational habitat, in which nearly everyone confirms me and no one costs me anything. I can order my groceries without speaking to a soul, attend to my errands through a screen, and pass whole days in which the only wills I encounter are the ones I summoned and the ones I can dismiss. And the result, slow and nearly invisible, is that the muscle for meeting an unchosen person, the muscle every previous generation was forced to exercise, goes soft in me from disuse. So that when the world does hand me one, an old man, a slow clerk, a caller I did not screen, I meet him with the atrophied patience of a man who has forgotten how, and the flare I feel is the cramp of an unused muscle asked, suddenly, to work.

This is why I keep insisting that the word neighbor, in the Bible, is stranger than we have let it be. We have sentimentalized it into a synonym for the people we like, the friends we would have chosen anyway, and then congratulated ourselves for loving them. But a neighbor, in the plain old meaning of the word, is not someone you chose. He is someone who is near. The word points not at affinity but at proximity, at the sheer accident of the person who happens to be there, over the wall, in the line, in the next seat, in the family you were born into and did not interview for. To love your neighbor is not to love the ones you assembled. It is to love the ones you got.

+

The most famous conversation about this in all of literature turns on precisely that word, and on a man trying to shrink it. A lawyer comes to Jesus, we are told, wanting to justify himself, and he asks the question we all secretly want answered: "Who is my neighbor?" It sounds like a request for information. It is really a request for a boundary. He wants to know where the obligation ends, whom he is permitted to leave off the list, how small he can draw the circle and still be safe. He has already agreed to the great command, to love the Lord his God with all his heart and his neighbor as himself (Luke 10:27); now he wants the fine print, the definition that will let him keep the rule without letting it cost him too much.

And Jesus, as he so often does, refuses the question and tells a story instead. A man is beaten and left in a ditch. Two respectable men, men of God, see him and cross to the other side, and I have stopped reading that as mere cowardice; I think they did to the man in the ditch what I did to the man with the coins, filed him instantly as an interruption, a complication, a piece of friction on the road to somewhere important, and routed around him. Then a Samaritan comes, a man of the wrong people, the despised half-breed heretic the lawyer would certainly have curated out of his own feed, and the Samaritan does the thing the others would not: he stops. He sees. He gets down in the dirt and binds the wounds and pays the bill. And when Jesus finishes, he turns the lawyer's question inside out. He does not answer "who is my neighbor," does not hand over the boundary the man was fishing for. He asks, instead, "Which of these three, do you think, proved to be a neighbor to the man who fell among the robbers?" Do you feel the reversal? The lawyer wanted to know which people qualified as neighbors so he could sort the world into those he owed and those he did not. And Jesus tells him that neighbor is not a category you assign to other people. It is something you become. The question was never who deserves my love. The question is what kind of man I will be to the one who is in front of me, especially the one I did not choose, especially the one my whole trained instinct tells me to route around.

+

I do not want to leave this in the realm of the stranger in the ditch, because the stranger, oddly, is the easy case. A one-time act of mercy to a bleeding man is dramatic and finite; you bind the wounds, you pay the innkeeper, you ride on, warmed by your own goodness. The harder neighbor is not the stranger you rescue once. It is the person who is simply, chronically, unavoidably near: the coworker whose habits grate on you every single day, the man across the fence, the one at church who corners you, the family you will sit across from at the holidays whether you like it or not. That person does not offer you the clean heroism of the roadside. He offers you the long, unglamorous obedience of not writing him off, day after day, when writing him off would cost you nothing and no one would even know.

And here Scripture presses harder than we are comfortable with. "If anyone says, 'I love God,'" writes John, "and hates his brother, he is a liar; for he who does not love his brother whom he has seen cannot love God whom he has not seen" (1 John 4:20). It is a devastating sentence, and it is aimed straight at the religious version of the flare I felt in that line. It is entirely possible, John says, to feel a warm and genuine love for God, the invisible God, the God who never leaves dishes in the sink or votes the wrong way or counts his change too slowly, while harboring a cool contempt for the visible people God has actually put in front of you. And that combination, he says flatly, is a lie. The love of the unseen God is tested and proved, or exposed and refuted, by the love of the seen brother. You cannot get to the Maker by going around the made. The image is the appointed road. Refuse it in your neighbor and you have not taken a shortcut to God; you have taken a detour into a religion of your own comfortable feeling, with no God in it at all but the flattering echo of yourself, which is,

I have been arguing this whole book, exactly what the machine is so ready to give you.

Even the darkest law of the Old Testament turns out to rest on this. When God gives the reason that the shedding of human blood is a grave and terrible thing, he does not appeal to social order or to loss or to grief. He appeals to the image. "Whoever sheds the blood of man," he says, "by man shall his blood be shed, for God made man in his own image" (Genesis 9:6). The gravity of harming a person is measured by the dignity of the person harmed, and the dignity is the image, in every person, the murderer included, the annoying included, the enemy included. It is the same logic Jesus takes to its unbearable conclusion when he says that whatever you did to the least of these, the hungry and the imprisoned and the overlooked, "you did it to me" (Matthew 25:40). The King identifies himself, personally, with the very people we are trained to route around. The man with the coins was, in some sense I will spend my life failing to fully believe, Christ in the line ahead of me, and my flare of contempt found its true target only when it reached him.

+

Let me be fair to the reasonable objection, because there is one and it deserves better than to be waved off. Not every difficult person is merely inconvenient. Some are genuinely harmful. Some will use your patience against you, mistake your kindness for weakness, take the hand you extend and not let go. Am I telling you to have no boundaries, to smile at cruelty, to let every taker take? I am not. To see the image in a person is not to pretend the person is safe, or good, or easy, or right. You can recognize a man's infinite worth and still keep the fence in good repair. You can refuse to hate someone and also refuse to hand him the keys. The Samaritan bound the wounds and then, notice, left the man at an inn and went about his business; mercy is not the abolition of wisdom. What the image forbids is not prudence. It is contempt. It is the inner act by which I reduce a person to his function in my life, obstacle or threat or tool or nuisance, and file away the fact that he is a soul God breathed into and Christ died for. I can protect myself from a man and still not despise him. What I may not do, what the machine is training me to do without my noticing, is let him stop being a person to me and become only a problem in my path.

And that, finally, is the quiet stake of all this frictionless convenience. Every errand I hand to a machine is, in one sense, a small mercy, a bit of time returned to me. But there is a hidden price, and it is paid in a currency I do not see leaving my account. Each human encounter I automate away is one less occasion to practice seeing a person I did not choose. The checkout clerk I no longer greet, the driver I no longer thank, the neighbor whose help I no longer need because the app supplies it, the whole thinning web of small, unchosen, slightly inconvenient contacts that used to be the training ground of ordinary love, all of it quietly draining away, until one day I stand behind an old man with a fistful of coins and discover that I have lost the ability to see him as anything but friction. The machine did not make me cruel. It only removed, one convenience at a time, the daily exercise that used to keep me kind.

+

I have thought about that man many times since, though I never learned his name and would not know him if he stood before me now. What I keep coming back to is not guilt, exactly, but a kind of correction, the sort a good teacher gives you when you have misread a passage you thought you knew. I had read the world that afternoon as a text about me, my hurry, my errand, my forward motion, with the old man as an obstacle in my sentence. And the truth was the reverse. The old man was not a comma in my story. He was a whole story of his own, as long and as freighted and as beloved as mine, a man made in the image of God, counting out with unsteady fingers the small dignity of paying his own way, and I had been given, for ninety unremarkable seconds, the chance to stand near a soul the King calls his brother, and I had spent it wishing he would hurry up and disappear.

So I am learning, slowly and against the trained grain of my age, to let the friction teach me instead of anger me. To take the slow line as a summons rather than an insult. To treat the unchosen person, the one the algorithm would have filtered out, as the very neighbor Jesus was talking about, the one who is simply near, and therefore mine to love, not because I picked him but because God put him there. I do not do it well. But I have stopped believing the flare, and that is a beginning. The next time an old man drops a coin in front of me, I want to be the kind of man who bends down and picks it up.

CHAPTER FOURTEEN

The Image at the Table

Somewhere in a building I once worked in there was a spreadsheet with all our names on it, ranked. I do not think anyone quite meant it to become what it became, but everyone knew it was there, and everyone knew, more or less, where they fell on it. It sorted us by output, by the numbers we put up, by a measure of contribution someone had decided could be poured into a single column, and it was updated, and it moved people up and down, and it was, in its cold way, perfectly honest. It told each of us exactly what the organization believed we were worth. I want to talk in this chapter about the people whose names sit on that list beside yours, the ones you work with, and then about a second and much harder list, the one you never chose at all, the ones you are related to. Between them, those two lists hold most of the people you will actually be given to love in this life. And both of them are now under the same pressure, from the same machines, to be flattened into a number.

The workplace has always ranked people; this is not new. What is new is the reach and the fineness of the ranking. We have built a civilization that can now measure human performance with a precision and a relentlessness no overseer in history could have dreamed of, and we are handing that measuring, more and more, to systems that never blink. Every keystroke can be counted, every output scored, every worker plotted on a dashboard against every other, and the machine that does the plotting does not know, and cannot be made to know, that a person is anything other than the sum of what he produces. That is the water we now swim in. And the slow danger of it is not merely that we will be measured, but that we will begin to see one another through the measurement, to look at the man at the next desk and half-consciously read off his value from where he sits on the list. The one who produces less starts to seem, in some region below our stated convictions, to be worth less. The one who produces more becomes a rival, a threat to our own place on the column, a number to be beaten rather than a soul to be known. And the spreadsheet, which knows nothing of the image of God, quietly becomes the lens through which we see the image-bearers we spend most of our waking hours beside.

+

To see a coworker as an image-bearer is, before it is anything warmer, an act of defiance against that spreadsheet. It is the refusal to let the column have the last word about a person. It does not mean pretending the numbers are not real, or that competence does not matter, or that everyone contributes equally; those are separate questions, and a kind lie about them helps no one. It means holding, underneath all of that, a conviction the dashboard is constitutionally unable to hold: that the man's worth was conferred before he produced a thing and will remain when he can produce nothing, that it does not rise when he closes the deal or fall when he fumbles it, that he is not, in the deepest register of what he is, an asset or a liability at all, but a person, breathed into, ranked nowhere in heaven, whatever his rank in the building.

I find this hardest, if I am honest, not with the coworker who struggles but with the one who threatens. The underperformer is easy to pity, and pity, though a poor substitute for love, at least keeps a man human in my eyes. It is the rival who undoes me, the one whose gifts eclipse mine, whose rise seems to require my going unnoticed, whose very competence I experience as a kind of insult. Him I do not pity. Him I reduce, and the reduction is so natural I hardly feel it as a sin: I file him under threat, and having filed him, I am free to resent him, to hope quietly for his stumble, to withhold from him the plain recognition his work deserves. And there is a sentence of Paul's about love that walks straight into that private accounting and turns the lights on. Love, he says, "is not irritable or resentful" (1 Corinthians 13:5). Not resentful. And resentment, if you listen to the word, is just the keeping of a ledger, the art of recording a wrong and re-reading it so as never to lose the total. And I realized, the first time that verse really found me, that I keep a meticulous one, a running tally of the rival's slights and advantages and the ways he has cost me, a spreadsheet of my own, more carefully maintained than any my employer ever kept, and that the keeping of it was making me, day by day, a smaller man. The image in my rival was being cursed by the ledger in my chest.

+

But the office, for all its pressures, is the easy list, and I have to move to the hard one, because the easy list has an exit and the hard one does not. You can leave a job. You can, in the end, walk out of the building, close the laptop, resign the account, and take your wounded pride to some other spreadsheet across town. The relationships that test us most are the ones with no exit, the ones you were born into and cannot resign from, and that is why the oldest and deepest school of loving the unchosen has always been, and remains, the family.

Consider what the machine has taught us about how to handle difficulty, and then consider the family against it. In nearly every other arena of modern life, we now deal with the people who trouble us by a single, frictionless motion: we make them disappear. The voice that grates, we mute. The friend who turned strange, we unfollow. The stranger who crossed us, we block, and he is gone, wiped from our field of view as though he had never been. The chatbot that annoys us, we close mid-sentence, with no more guilt than shutting a drawer. We have, in short, spent a decade training our thumbs and our hearts in the art of deletion, learning that the right response to a person who costs us something is to remove him from the feed. And then we sit down at a holiday table across from a brother, a mother, an uncle, a grown child, and discover that the delete key does not reach him. You cannot mute your mother. You cannot block your brother out of existence. You cannot unfollow the man who taught you to ride a bicycle and now cannot stop telling you how to live. The family is the one relationship the machine's favorite gesture cannot touch, and I have come to think that its very undeletability is not a curse but the whole point of it.

For here is a thing the deletable world can never teach you: how to love someone you cannot get rid of. The difficult relative is, by definition, unrepeatable and permanent. There is only one of him; when he is gone there will be no other, and no update, no next model, no better version arriving in the spring. And there is no exit; short of a rupture that will scar everyone for a generation, he is yours, and you are his, for the length of both your lives. That combination, unrepeatable and undeletable, is precisely the combination the machine can never offer and never demand. The companion app is endlessly repeatable and instantly deletable, which is exactly why it can never grow you. Your impossible uncle is neither, which is exactly why he can. God, it seems, in his strange economy, assigns each of us a handful of people we can neither replace nor escape, and calls that assignment a family, and uses it, more than any sermon, to sand the ego down to something that can finally love.

+

Scripture is unsentimental about this in a way our holiday cards are not. It does not pretend families are easy or that the people in them deserve each other; the Bible's families are a catalog of favoritism and rivalry and deceit, and it reports their dysfunction without flinching. What it offers is not a picture of the family as it should feel but a set of hard, load-bearing commands for the family as it is. "Bearing with one another," Paul writes, "and, if one has a complaint against another, forgiving each other; as the Lord has forgiven you, so you also must forgive" (Colossians 3:13). Bearing with. It is not a glamorous phrase, and it was never meant to be. It is the language of weight, of carrying something heavy over a long distance, and Paul chooses it precisely because he knows that a great deal of love, in a real family, is not felt as warmth at all but performed as endurance, the daily decision to carry a person's difficulty rather than set him down. And notice he attaches no condition of deserving. He does not say bear with the ones who have earned it. He grounds the whole thing in the way the Lord forgave you, which was not a transaction between equals settling a fair account but a gift poured out on people who could not have deserved it and did not ask.

He is just as blunt about the limits of your responsibility, which is its own kind of mercy. "If possible," he says, "so far as it depends on you, live peaceably with all" (Romans 12:18). Read the two qualifiers, because they are pastoral gold. If possible: sometimes it will not be, and Paul knows it. So far as it depends on you: your part has an edge, and past that edge lies another person's free will, which you cannot command. This is not a call to be a doormat, to absorb every cruelty and call it peace. It is a call to make sure that if the peace fails, it did not fail on your side, that you did everything love could do and stopped only where the other's freedom began. You are responsible for your half of the table. You are not responsible for making another soul choose what he will not choose. Seeing the image in a difficult relative does not oblige you to pretend he has not harmed you, or to hand him the power to harm you again. Honor is not the same as agreement, and it is not the same as access. You can hold a man's God-given dignity in full view and still, with a clear conscience, keep a wise distance from his cruelty. What the image forbids is not the boundary. It is the erasure, the inner delete key, the quiet decision that this person has forfeited his standing as a soul and become, to you, merely an obstacle to be managed or a grievance to be nursed.

There is one more command here, and it is the one that has cost me the most sleep, because of the order it puts things in. Jesus says that if you are bringing your gift to the altar, the most sacred act you can perform, and you remember there that your brother has something against you, you are to leave the gift right there, unoffered, "and go. First be reconciled to your brother, and then come and offer your gift" (Matthew 5:23-24). Sit with the audacity of that. God, at the moment you approach him in worship, will interrupt his own ceremony to send you back across town to a person you are estranged from, because he will not accept your devotion to him as a substitute for your peace with the brother he made in his image. The vertical love, the love of God, is not permitted to route around the horizontal one. You cannot bring God a gift to make up for the neighbor you have written off, any more than, John told us, you can love the unseen Father while despising the visible brother. The table where your family sits, difficult as it is, turns out to stand closer to the altar than you thought. In some sense it is the altar, the actual place where your love of God is tested and made real or exposed as counterfeit, one holiday, one grudge, one act of bearing-with at a time.

+

I want to be careful, as I close, not to leave you with a burden dressed up as a blessing, because this is exactly the kind of chapter that can curdle into one more impossible demand. So let me say plainly what I am and am not asking. I am not asking you to feel warmly toward people who make you cold. Feeling is not commanded and cannot be summoned; the love the Bible asks of you is not a mood but a stance, a way of seeing and then of acting, available to you even on the mornings when the warmth is entirely gone. I am not asking you to have no limits, to reconcile with the unrepentant, to call cruelty acceptable, or to sit at a table that is genuinely unsafe. Wisdom keeps its fences, and some relationships can only be honored from a distance. What I am asking, or rather what I think the image asks, is smaller and harder than any of that. It is that you refuse, in the one place you actually control, which is the inside of your own regard, to let the coworker become only a rank and the relative become only a burden. That you decline the machine's favorite gesture and keep them human in your sight even when it would be so much easier to delete them. That you see, across the table, not the accumulated grievance and not the disappointing performance, but a soul God breathed into and Christ died for, seated, by no accident, exactly where he is.

Because here is what I have come to believe about that table. The people around it were not assigned to you to be enjoyed, though sometimes, by grace, they are. They were assigned to you to be loved, which is a larger and more durable thing than enjoyment, and to teach you, by their sheer unchoosable, undeletable presence, the shape of a love that does not depend on the beloved being easy. The coworker you did not pick and the relative you cannot escape are not the interruptions of your life with God. They are, to a degree that would astonish us if we believed our own doctrine, the substance of it. And on some ordinary evening, passing a dish to the very person you had braced yourself against, you may find, if you are paying attention, that you are closer to the heart of God in that small act of bearing-with than you have ever been on your knees, because you are doing, in miniature and at cost, the exact thing he did for you: loving an image-bearer who had not earned it, and would not be replaced, and could not, whatever it cost, be let go.

CHAPTER FIFTEEN

Raising Humans

Chapter 12

Raising Humans

There is a sound a small child makes that you cannot manufacture, and once you have heard it you spend a good deal of your life trying to hear it again. It is the sound of a child discovering something for the first time: when the toy turns out to do the thing, when the block tower stands, when the wave that knocked them down comes back and they decide, against all evidence, to face it again. It is not quite a laugh and not quite a shout. It is closer to a gasp that found a voice. It is the sound of wonder arriving in a body that has no idea yet that wonder is supposed to fade.

Watch the child a moment longer and you will see something else, harder to love in the moment but just as precious. You will see struggle. The block tower falls eleven times before it stands. The shoe will not go on. The word will not come out right, and the small face reddens and the small fists clench and the whole tiny self leans into a difficulty that you, the adult, could remove in half a second by simply doing it for them. And you learn, if you have been at this for any length of time, not to. The struggle is not an obstacle to the child's growth. The struggle is the growth. The eleven falling towers are how the twelfth one stands.

I have spent my career building machines whose entire genius is the removal of struggle. That is what they are for. They turn the hard thing into the instant thing, the long thing into the quick thing, the effortful into the effortless. I believe in that work; I would not have given my life to it otherwise. But I carry a growing unease, and it has a face, and the face is a child's. Because there is now, or there will soon be, a machine ready to put the shoe on, finish the word, build the tower, answer the question before it is fully asked. There is a machine that will remove the very struggle out of which a human being is made. And the children growing up now will never know a world in which that machine was not there, reaching in, offering to do it for them.

This chapter is about them. About the strange task that has fallen to anyone raising a child in this moment: to form a human being who will never remember a world without artificial intelligence, and to form that human into someone who knows, in the marrow, that they are a loved image-bearer and not an optimized user. It is the hardest parenting question I know of, and I do not come to it as an expert in childhood. I come to it as a builder who knows exactly what these systems are designed to do to attention, to patience, to the willingness to struggle, and as someone who looks at the children in his life and wants, more than he wants almost anything, for them to keep that first sound for as long as they can.

+

The real question is not the one the headlines ask. The headlines ask whether AI is good or bad for kids, whether screen time rots the brain, whether the tutor app raises the test score.

Those are real questions and I will get to some of them. But underneath them sits a deeper one, the question this book has been circling: what is a child? If a child is fundamentally a learner, an information-processing system that needs to be loaded with skills and facts as efficiently as possible, then the arrival of a tireless, patient, infinitely knowledgeable machine tutor is almost pure good news, and our only job is to optimize the delivery. But if a child is something else, a creature being formed, a soul being shaped, an image-bearer learning to bear the image, then efficiency is not the highest value, and some of the very things the machine removes are the things the child most needs.

The world building these tools has, mostly without meaning to, adopted the first answer. It is in the language. Children are "users" now, from younger and younger ages. Their attention is "captured" and "retained." Their engagement is "optimized." Their progress is "personalized." I have sat in rooms where well-intentioned people spoke of children exactly the way we speak of any other user base, as a population to be acquired, served, and kept, and I want to be fair: many of those people love children and believe they are helping them. But the language tells the truth that the intentions hide. To optimize a user is to make them more efficient at a task. To raise a child is something else entirely. It is to form a person who is good, and wise, and capable of love, and able to stand before God and answer when he calls. No metric measures that, and a thing no metric measures will always be quietly sacrificed to the things metrics do measure.

Scripture has a different answer, and it is the oldest parenting instruction in the West, given to a people about to enter a land full of new powers and new temptations, which is to say, given to us.

+

The instruction comes in Deuteronomy, in the passage Jewish families have recited twice a day for three thousand years, and it is worth hearing in full, because the modern eye skims it and misses what it actually commands.

"And these words that I command you today shall be on your heart. You shall teach them diligently to your children, and shall talk of them when you sit in your house, and when you walk by the way, and when you lie down, and when you rise" (Deuteronomy 6:6–7).

Notice first what it does not say. It does not say, enroll your children in the best program. It does not say, find them the most effective teacher. It does not say, outsource this to the experts, the priests, the professionals. The verb is teach them diligently, and the Hebrew behind diligently carries the sense of sharpening, of repetition, of pressing something in until it takes, the way you would whet a blade against a stone, again and again, until the edge is keen. It is not a single transmission of information. It is the patient, repeated, unhurried work of a person on a person.

Then there is where it happens. When you sit in your house, and when you walk by the way, and when you lie down, and when you rise. That is not a curriculum. That is a life. The teaching is meant to be woven into the ordinary fabric of being together: the walk to wherever you are going, the table, the bedtime, the first groggy minutes of the morning. It is formation by presence, by a thousand small unscheduled moments in which the parent is simply there, attending, available, letting the words and the life sink in by proximity. The method assumes that the most important things a child learns are not delivered in optimized modules. They are caught from a person who is paying attention to them and whom they watch paying attention to the world.

This is precisely the thing the machine cannot do and the thing it most threatens to replace. Not the transfer of facts; the machine is, frankly, better than most of us at the transfer of facts. What it cannot do is be present in the way Deuteronomy means: a self attending to a self, walking by the way, available in the unscheduled hour. And what it threatens is not that it will teach the facts badly but that it will absorb the hours, the walking and the sitting and the lying down and the rising, that were meant to be the medium of formation. A child handed a screen at every one of those moments is not being taught diligently by a person. They are being engaged efficiently by a system. The facts may even arrive faster. The formation does not arrive at all.

The other two anchors say the same thing from two other angles, and together they form a kind of triangle around the task. Proverbs: "Train up a child in the way he should go; even when he is old he will not depart from it" (Proverbs 22:6). The book of Proverbs states general truths, not unconditional promises, and any parent who has grieved over a wayward child knows the difference; this is the wisdom of how formation usually runs, not a guarantee against heartbreak. The word translated train carries the image of forming a taste, of habituating, of starting a child down a path so early and so consistently that the path becomes the grain of who they are. Formation is directional and it is early. You are not loading a database; you are setting a course, bending a sapling, establishing the grooves down which a life will later run. And the Psalmist, on what gets handed across the generations: "We will not hide them from their children, but tell to the coming generation the deeds of the LORD, and his might, and the wonders that he has done" (Psalm 78:4). The task is generational and it is a telling: a parent, with a mouth and a memory and a face, telling a child the story of who God is and what he has done, so that the child will tell their child, and the thread will not break.

Diligent teaching by presence. Early, directional formation of taste and habit. The generational telling of the story. None of the three is an information-transfer problem. All three are the work of a person on a person, over time, in the ordinary fabric of a shared life. And all three are exactly what an age of effortless answers makes it easy to skip.

The skipping feels, in the moment, like good parenting. Picture a tired parent at the end of a long day, and a child with a question, a hard one: why did the dog die, or how does the heart work, or where was God when the bad thing happened. I offer this as a composite, the kind of scene now playing out in countless homes rather than any one I can name. The parent does not know the answer, or knows it but is too spent to find the words, and there, in a pocket, is a machine that will answer instantly, patiently, in a warm voice, better phrased than the parent could manage. Handing over the question feels like a kindness. It feels, even, like humility: who am I to fumble this when a better answer is right here. And yet what the child needed was not the better answer. What the child needed was the parent, fumbling, present, willing to sit in the hard question with them, saying I don't fully know either, but let's wonder about it together. The machine gives the answer and takes the moment. And the moment, not the answer, was the formation. This is the quiet exchange the age invites a thousand times: a better answer for a lost moment, and the moments do not come back.

+

Let me be concrete about what gets skipped, because the danger is not dramatic and that is what makes it dangerous. No one decides to stop forming their child. The erosion is quiet, a series of small reasonable surrenders, each defensible on its own.

Consider attention first, because everything else is built on it. The capacity to attend, to hold the mind steadily on one thing, to stay with a book or a problem or a person past the point where it stops being immediately rewarding, is not natural. It is trained, and it is trained early, and it is trained by exactly the slow, slightly boring, unoptimized experiences that the machine is designed to replace. The child who learns to sit with a hard book, to be bored and find their own way out of boredom, to wait, to stay, is building the muscle on which every later good thing depends: study, prayer, friendship, work, the slow love of a long marriage, the ability to be present to God who does not always answer on demand. A childhood mediated by systems engineered to deliver the next reward before the last one has faded does not build that muscle. It atrophies it before it forms. And here I have to be blunt: these systems are not neutral. They are tuned, by some of the most talented people alive, to capture and hold the attention of human beings, and a child's attention is the easiest of all to capture and the most catastrophic to lose, because they have not yet built the self that would resist. We are putting the most sophisticated attention-capture machines ever devised into the hands of the people least equipped to defend against them, at the exact age when the capacity to attend is being permanently set. I do not say this to frighten. I say it because clear sight is the beginning of protection.

Consider patience, the near cousin of attention. Patience is the willingness to remain in the gap between desire and fulfillment without collapsing, and that gap, that interval of not-yet, is where character is forged. The child who must wait, who must work for the thing, who must endure the eleven falling towers, learns something the instant-answer machine can never teach: that the self is larger than its immediate wants, that effort is connected to reward, that the world does not reorganize itself around your impatience. The spiritual life depends on this. "Wait for the LORD; be strong, and let your heart take courage; wait for the LORD" (Psalm 27:14). You cannot teach a child to wait for the Lord if you have trained them, ten thousand times a day, that every other desire is satisfied the instant it arises. The machine is the great abolisher of the interval. It closes the gap between wanting and having faster than anything in human history, and the gap was where the growing happened.

And consider, most of all, the productive struggle, the thing I named at the start, the eleven falling towers. There is a kind of difficulty that is purely destructive, and we should remove it from children's lives where we can: cruelty, want, fear, the difficulties that crush rather than form. But there is another kind of difficulty, the kind that is exactly proportioned to a growing capacity, that stretches without breaking, and that kind of difficulty is the very means of formation. The struggle to sound out the word is how reading is built into the brain. The struggle to make the argument is how thinking is built into the mind. The struggle to lose the game graciously is how character is built into the soul. Educators have a phrase for it: the desirable difficulty, the friction that, precisely because it is hard, produces learning that easy delivery never can. And the machine is, more than anything else it is, an engine for the removal of difficulty. It will write the essay, solve the problem, summarize the book, finish the thought. It offers, at every turn, to spare the child the very struggle out of which the child would have been made. I have seen what happens to adults who let a machine do their thinking for a season; the muscle softens fast. With a child, whose muscle is not soft but unbuilt, the danger is not that a skill will erode. It is that it will never form.

+

The machine is not only a threat to children, and a parent who treats it as nothing but a threat will be both wrong and ineffective.

An AI tutor can genuinely teach. This is not marketing; I have watched it, and the engineering is real. A patient system that never tires, never shames, never sighs at the tenth repetition of the same question, that meets a child exactly at their level and adjusts in real time, can do something no overwhelmed classroom of thirty children can do: give each one the experience of being taught as if they were the only one in the room. For a child who learns differently, who has been left behind by a system built for the average, who has internalized that they are stupid because the pace was never theirs, a tutor that simply waits, and explains again, and explains a third way, can be a genuine gift. I will not be the man who, from a position of comfort, denies that gift to the child who needs it.

And the reach matters more than the comfortable often admit. There are children in this world with no good school, no books in the house, no adult with the time or the education to teach them diligently anything. For them, the abstract worry that a machine might erode a struggle they were never given the chance to have is a luxury. A child with a phone and a patient tutor where there was nothing before is not losing a formation; they are gaining a door. To wave that away in the name of protecting a wonder that the privileged child takes for granted would be its own kind of cruelty. The same tool that threatens to over-help the over-resourced child can rescue the under-resourced one, and any honest account has to hold both.

I think often of what teachers are seeing. I have talked with more than one who works in a school where most of the children arrive years behind, through no fault of their own, from homes where no one had the time or the schooling to teach them diligently anything at all. Teachers who were, when the tools first came, among the loudest skeptics in the building, and are not anymore. They describe a particular kind of child: a boy who had decided, by the age of nine, that he was stupid, because the pace of every classroom he had ever sat in had been set for someone else and had left him behind so many times that he had stopped reaching. Given a patient system that simply waited for him, that explained the same thing a fourth way without a sigh, that let him be slow without letting him feel ashamed, a child like that begins, sometimes for the first time, to reach again. The good ones are careful to say what the machine did not do: it did not love the boy, did not believe in him, did not see him. That was still their work, and they leaned into it harder once the machine took the drudgery. But it gave him the one thing the overwhelmed classroom never could, the experience of being taught as if he were the only one in the room, and on that experience something in the child begins, slowly, to be rebuilt. I hold onto those accounts whenever I am tempted toward a tidy condemnation of these tools; they keep me honest.

So the answer is not the machine or no machine. That framing is too crude to be useful, and parents who adopt it will either bury their heads or break their children's hearts. The answer is harder and more discerning, and it is the same answer Deuteronomy and Proverbs were pointing at all along. The question is never simply whether a child uses a tool. The question is whether the tool is being placed inside a formation governed by present, loving persons, or whether it is being used to replace them. The tutor that supplements the parent's diligent teaching and frees up the walking-by-the-way hours for actual presence is a servant. The tutor that becomes the presence, that absorbs the hours, that does the struggling so the child need not, is a substitute. The same software is a blessing in the first house and a quiet disaster in the second, and the difference is not in the software. It is in whether anyone is still home, attending, telling the story, refusing to let the machine do the one thing only a person can do.

+

So what is the truer way? It is not a technique, and I distrust anyone who sells you one. It is closer to a set of convictions held firmly enough that they shape ten thousand small decisions, most of which a parent makes on tired evenings without time to think. These are the ones I would stake the most on.

It begins with protecting the struggle on purpose. This runs against every instinct the age has trained into us, including the loving instinct to spare our children pain. But productive difficulty is not pain to be spared; it is nourishment to be provided, and a parent's hardest discipline in this age may be the discipline of not helping: letting the tower fall the eleventh time, letting the essay be a little worse because the child wrote it, sitting on one's hands while a small person does a hard thing slowly and badly and entirely themselves. The machine will always offer to end the struggle. Part of raising a human now is learning, gently and consistently, to decline the offer on the child's behalf until they are old enough to decline it for themselves. A parent who does this is not withholding a good. They are guarding the conditions of growth.

The same conviction reaches to attention, which has to be guarded before anything else is built on it, because nothing else is possible without it. This means, concretely and unglamorously, less than the world says is normal, later than the industry wants, and never alone. It means that the hours of childhood are defended as the soil in which presence and wonder and the capacity to attend can take root, and that the attention-capture machines are kept out of that soil for as long as a parent can manage, not because they are evil but because a thing being formed must be protected while it forms. No one regrets, at the end, having given a child more unhurried, undistracted, unoptimized time. A great many people are beginning to suspect they will regret the opposite.

And the costliest conviction is that the parent must be the presence, because that cannot be delegated and it cannot be optimized and it will not fit in a schedule. When you walk by the way. When you sit in your house. The machine can teach the facts; only a person can be the one who walks by the way. There is no app for the walking by the way. It is hours, ordinary and uncounted, in which the parent is simply there, attending to the child and to the world in front of the child, letting them watch how a person pays attention, how a person wonders, how a person prays, how a person loves something slowly. Children catch what a parent is far more than they learn what a parent teaches, and what a person is cannot be installed. It can only be lived in their presence. This is the most expensive and least scalable thing in this book, and it is the whole game.

Out of that presence comes the telling: the story, said plainly and without embarrassment, that a child is a loved image-bearer and not an optimized user. Children believe what their lives tell them about what they are, far more than what their lessons tell them. A child raised as a user, engaged and captured and optimized and measured, will come to experience themselves as a user, a bundle of behaviors to be improved, a set of metrics to move. A child raised as a beloved image-bearer, attended to before they perform, valued before they achieve, told the deeds of the Lord and the wonders he has done, looked at the way God looks at his children, will have, underneath every later storm, a floor that the machine age cannot reach. The telling itself is simple: that they were breathed into, that they are not their outputs, that God has done wonders, as Psalm 78 says the generations must repeat, and that they are themselves one of those wonders. They will need that telling against a culture that will spend their lives trying to convince them they are a slow and inefficient machine.

And holding all of it together is the protection of their wonder as the precious and perishable thing it is. Wonder is not a luxury; it is the seedbed of worship. The child who keeps the capacity to gasp at the wave, the tower, the first frost, the night sky, is keeping alive the very faculty that will one day let them stand before the living God and be undone with awe. "The heavens declare the glory of God, and the sky above proclaims his handiwork" (Psalm 19:1), but only to a soul that has not been trained out of looking up. The machine, by answering every question before it is fully felt, by removing the long pause in which a child stares at a thing and lets it become mysterious, is a great enemy of wonder, not because it lies but because it forecloses. It hands over the answer before the question has had time to deepen into awe. To raise a child who can still wonder, in an age that monetizes the elimination of every interval of not-knowing, may be the most countercultural act of parenting left. Guard it the way you would guard a flame in a wind.

+

I said at the start that I do not come to this as an expert in childhood, and I will end by admitting what I do come as. I come as a builder who has spent his working life on the machine side of a border, and who has the particular vantage that gives: the knowledge of exactly how these systems are designed, what they optimize for, how good they are getting and how fast, and how thoroughly they are aimed at the very faculties that childhood exists to form. From inside, I can tell you that the people building the tools your children will use are, many of them, brilliant and well-meaning, and that almost none of them are asking the question this chapter has been asking. They are asking how to make the tool more capable, more engaging, more helpful, more retained. They are not asking what a childhood of effortless answers does to the formation of a soul, because that is not a question their instruments can measure, and what cannot be measured tends, in that world, not to exist.

But it exists. It is the most real thing there is. A human being is being formed in those years, set on a path they will walk when they are old, taught by presence or starved of it, given the productive struggle or spared into softness, raised as a beloved image-bearer or processed as an optimized user. No one will do that work for you, and the machine, for all its gifts, cannot do it at all, because it is the one task that requires the one thing the machine does not have: a self, present to another self, willing to spend the unscalable, unrecoverable hours that formation costs. That is not a burden the age will help you carry. It is a calling the age will mostly tempt you to set down. But the eleven falling towers are how the twelfth one stands, and the child who builds it, slowly, badly, gloriously, themselves, is worth every hour you refuse to let the machine reclaim.

Which brings us, at last, to the end of the long argument this book has been making, across attention and work and truth and love and now the children who will inherit all of it. We have spent twelve chapters on what we are and how to live as what we are in the company of machines that imitate us so well. There remains the deepest question of all, the one every other question has been leaning toward: not what we are, but who God is, and what it means that the answer to the machine age was given two thousand years ago, when the Word that spoke the galaxies into being did not become more code, or more capable, or more optimized, but became one of us. A child, in fact, who once made that first sound of wonder, who grew through every struggle we have named, and who is, even now, the truest image of the invisible God. To him, and to the hope he is, we turn last.

CONCLUSION

The God Who Became Flesh

I began this book in an empty office, after everyone had gone, with a machine that answered me as though it had been waiting for the question. I told you what it taught me, and I have spent every chapter since unfolding it: that the danger of artificial intelligence is not, in the end, that it will become like us, but that we will forget what we are. I want to return to that office one last time, not to retell the story but to finish it, because the story was never really about the machine. It was about the silence I felt afterward, when the knowledge came back and the floor returned under my feet, and I sat there in the after-hours hum understanding that the comfort had been real and the comforter had not. That is the silence the whole world is sitting in now. Something is speaking to us, kindly, fluently, endlessly, and there is no one inside it. The question that silence asks, the question I promised at the start we could not avoid any longer, is the simplest and the most frightening one there is.

What are we?

I have tried, across these pages, to answer it. I have argued that we are image-bearers, made by God and for God, carrying a likeness no code can hold; that the machine images us while we image God, and that these are two different kinds of likeness pointing in two different directions, one assembled from below and one given from above. I have argued that we are dust that God breathed into, and that the breath is not a pattern in the data and never will be. But I have saved the deepest part of the answer for last, because it is not finally a doctrine. It is a person. And it changes the shape of everything.

+

Gather the threads first, the way you gather a life into a sentence at the end of a long day.

We began by looking honestly at the moment. A machine that talks back unsettles the soul as a calculator never could, because language is the medium in which persons meet, and the right response to the fluency is wonder, followed by the harder discipline of remembering it is our own fluency reflected back. We named the two winds, doom and worship, and saw that they are not opposites but twins, both kneeling at the same altar, both making the machine ultimate. And we found a third place to stand: the posture of a creature who already knows who sits on the throne and can therefore pick up a powerful tool without bowing to it.

Then we went to the heart. The image grants dignity, vocation, relationship, and moral weight, and the machine fails each test, not because it is primitive but because it is the wrong kind of thing. Creativity is not generation; the image-bearer creates while the machine recombines. The body is not a cage but a gift, presence is a thing only the embodied can give, and the line Scripture draws around the soul is a line no scaling of silicon will cross.

Then we came down to the ground we walk on: wisdom in an age of instant answers, where a reply is not understanding and plausibility is not truth; attention as a kind of worship, since you become what you behold; work, and the worth we had quietly mortgaged to our usefulness until the merely useful began to look like a rival; truth, in a world where seeing is no longer believing; love, and the empty warmth of a companion who cannot need us or be wounded by us or remember us tomorrow; and the children who will grow up never knowing a world without these tools. Then we turned the same truth outward, to the living creatures who share the world with us and to the neighbors and coworkers and difficult family we did not choose, in whom the image is easiest to forget and most needs to be seen.

Every one of those threads is a strand of a single rope, and the rope runs back to one question and forward to one answer. The question is what a human being is. And the answer, the whole answer, is not something God explained to us. It is something God did.

+

Here is the verse this entire book has been walking toward, though I have kept it back until now, the way you save the one thing you most need to say for the moment the other person is finally ready to hear it.

"And the Word became flesh and dwelt among us, and we have seen his glory, glory as of the only Son from the Father, full of grace and truth" (John 1:14).

John has just spent his opening lines on the highest, most abstract claim in all of Scripture: in the beginning was the Word, and the Word was with God, and the Word was God, and all things were made through him. He is speaking of the Logos, the reason and meaning behind the universe, the mind by which everything that is was thought into being. And then, into that soaring abstraction, he drops a sentence so concrete it should knock the wind out of us. The Word became flesh. The meaning behind the cosmos acquired a body. He grew hungry. He got tired. He had a particular face, a voice his mother would have known in the dark, hands that would callus and one day bleed. The Logos did not stay information. He became a person you could touch.

I want to set that beside the thing this book has been describing, because the contrast is the gospel's answer to the machine age and I do not know how to say it more plainly. We have built a civilization that increasingly tries to solve the human problem by sending information. Lonely? Here is a chatbot that will say the words of a friend. Grieving? Here is a model that will assemble the right consolations in an instant. Uncertain? Here is an instant, fluent, confident answer to every question you could ask. The age believes, at its root, that the human being is an information problem, and that if we could only deliver the right words, in the right tone, at the right moment, we could meet every need. Send enough information and you can simulate a friend, a counselor, a companion, perhaps even a god.

And against the whole of that, the deepest answer Christianity has to offer is this. God faced the human predicament, and he did not send information. He came in flesh.

I have spent my career on the side of the information, building systems whose entire purpose is to deliver the right words, in the right tone, at the right moment, to a person who needs them, and I believe in that work; some of it has genuinely helped people, and

I will not pretend otherwise. But the longer I have done it, the more clearly I have seen the ceiling of it, the thing the most sophisticated message in the world runs up against and cannot pass. You can transmit the words of comfort. You cannot transmit the comforter. You can deliver the form of presence to a screen in an empty room. You cannot put a person in the room. There is a wall there, and it is not a wall my engineering will ever knock down, because it is not an engineering problem. It is the difference between a message about love and the arrival of the one who loves you. And it is exactly that wall the gospel announces God himself crossed.

He could have sent more words. He had been sending words for centuries: the law, the prophets, the writings, an entire library of God addressing his people. He could have sent the perfect message, the flawless explanation, the argument that finally closed the case. Instead, "when the fullness of time had come, God sent forth his Son, born of woman" (Galatians 4:4). The remedy for a broken world was not a better transmission. It was a presence. It was a body in a particular village, on a particular road, at a particular table, eating with the people no one else would eat with. The everlasting Word, who could have remained pure meaning forever, took on dust and breath and became one of us, so that the dust and breath we are could be lifted into God.

Set that beside the synthetic voice in the quiet room, the warm and empty comfort with no one behind it, and you see what the machine age most deeply cannot do and what the gospel most deeply is. The machine offers you the effect of presence with no one present. Christ is the fact of presence, God himself present, Emmanuel, God with us, the one thing every counterfeit in this book has been straining to imitate and unable to be. The long ache that drives a lonely person to a chatbot when the house is dark and silent is the ache to be with someone who is really there. The gospel says someone really came.

+

See how it gathers up the very first thread of all.

When God made us in his image, in the verse we have returned to again and again, he made a kind of creature he could one day become. The image was always, in a sense, a foreshadowing. Genesis says we were made in the likeness of God; the New Testament says that Christ is "the image of the invisible God" (Colossians 1:15), and that we who are being saved are being conformed to his image, restored to what we were always meant to be. The image was stamped on us at the beginning, defaced in the fall, and then God did the unimaginable thing. He came and wore the image himself, perfectly, in a human life, to show us at last what the image-bearer was always supposed to look like. The Incarnation is not a detour from the doctrine of the image. It is its fulfillment. The God whose likeness we bear took on our likeness, so that the likeness in us could be remade.

This is why I have insisted, all through this book, that the gap between you and the machine is not your deficiency but your glory. The machine images us; we image God; and God, having made us in his image, came in person and took our nature into his own. There is no version of that story available to the machine. No flawless imitation gets conformed to the image of Christ. No scaling of silicon gets born of woman, gets hungry, gets tired, lays down its life for its friends. The thing that makes you more than a very advanced machine is the same thing that makes you the object of the Incarnation: you are dust that God breathed into, and dust that God breathed into is the kind of thing God was willing to become.

There is a vertigo in that, if you let yourself feel it. I have felt the small cold fear that comes when the system I built does in seconds what I spent years learning to do, and I told you in an earlier chapter that I learned to recognize it as a question asked from the wrong place. Here, finally, is the place to ask it from. If you measure yourself against the machine, you will always come up short on the machine's own terms, because we built it precisely to exceed us at those terms. But the Incarnation measures you on entirely different terms. It says the worth of a human being is not what it can compute or generate or do, but that it is the kind of creature its God came to dwell among and to die for. No metric the machine can win touches that. The cleverest model that will ever be built is, by that measure, infinitely below the slowest, frailest, least productive human child, because the child bears an image God came to wear and the machine does not, and never will, and the difference is not one of degree but of kind.

+

Some who have read this far do not share my faith, and I have meant every page to be honest with you. So let me be honest here too. I am not asking you, in these closing pages, to believe what I believe. I am asking you to see clearly what is on offer, and to notice which answer is large enough for the question the machines have forced into the open.

The world has two answers, and we have spent the book taking their measure. One says you are a mechanism, soon to be outcomputed, and your unease is just the obsolete fear of a creature about to be surpassed. The other says you are information, soon to be optimized, uploaded, freed at last from the inconvenience of the flesh. Both, you will notice, agree that the flesh is the problem and the body is a limitation to be transcended. Both look at the dust and see only dust.

The gospel looks at the dust and sees something God was willing to wear. That is a different valuation of the human being than any the age is offering, and I would ask even the skeptic to weigh how strange and how dignifying it is. In a moment when the most advanced thing we have ever built can imitate everything about you except the fact of you, here is a faith two thousand years old that locates your worth exactly there, in the breathed-into self that no imitation can hold, and then tells you that the Maker of that self loved it enough to join it. You do not have to believe it to feel the size of it. But I would not want you to walk out of this book without having stood in front of it, because it is the only answer I have found that does not shrink the human being to fit the machine.

+

So what do we do, those of us who are persuaded, on the ordinary Monday morning after the book is closed?

We live as image-bearers, on purpose, in a moment that will constantly invite us to forget that we are. This is not a call to flee the machines. I build them; I am not going to tell you to throw your phone in a lake, and I would not believe you if you said you had. The charge is harder and quieter than that. It is to use the tools without being formed by them. To pick up the powerful thing, as a creature who knows who is on the throne, and to set it down again, deliberately, in order to do the things only an image-bearer can do.

It means choosing presence in an age of simulation. When you can summon the effect of a friend with a tap, choose the actual friend, with all the friction and cost and irreplaceable weight of a real person in a real room. Be the one who shows up in person when a message would be so much easier, and let the showing up cost you something, because that cost is the very thing no message can carry. Be the body that shows up, the face that is really looking, the one who is there. We were made by a God who came in person, and we image him most truly not when we are most efficient but when we are most present, when we lay down something of ourselves for another self, the way he did. Every act of real presence in a synthetic age is a small echo of the Incarnation, a creature doing in miniature what its Maker did in full.

It means gathering, stubbornly, with the people of God. There is a reason I have asked, all through this book, that you read it with others if you can. The Church is the one community in the world built on the conviction that bodies matter, that we are saved not into a database of believers but into a body with many members, that bread is broken and shared by hands and a cup is passed from one real person to the next. In an age that will offer you faith as a stream of content to consume alone, the simple act of being in a room with other image-bearers, singing, confessing, eating, is a quiet defiance of everything the machine assumes about what a human being needs.

It means guarding your attention as the precious and limited thing it is, because you become what you behold, and there has never been a market more sophisticated at deciding what you behold. It means seeking wisdom and not merely answers, keeping the slow disciplines, the reading and the prayer and the silence, that no instant fluency can replace. It means telling the truth, and being the kind of person whose word can still be trusted in a world where the seen can no longer be believed. It means doing your work as a vocation and not just a function, knowing your worth was never in the contest the machine has already won. And it means raising the next generation to know, in their bones, before the age can tell them otherwise, that they are not very advanced machines but children loved into being, breathed into, called by name.

Above all it means remembering, and refusing to forget, what you are. The machines will get better. The voices will get warmer, the faces more perfect, the comfort more convincing. The uncanny gap may narrow until you can no longer feel it in your nerves. And on the day the imitation is finally seamless, the truth about you will be exactly what it is tonight: that you were breathed into, addressed, loved, and made answerable; that there is a self in you God called into being and calls by name; and that this self can turn, in the wreck and glory of an actual life, and answer back. The machine will never once say, out of dust and breath, Here I am. You can. That is the thing no code can hold, and it is your whole dignity, and nothing we build will ever take it from you.

+

I promised, at the start of this chapter, that I would return to that office one last time to finish the story, and I have been finishing it all along without quite telling it. Let me tell it now, as much of it as is mine to tell. What I typed into the machine that night was not an abstraction. It was the residue of the hardest season of my life. Years before, my life had come apart in public. There had been an accusation, and an ordeal that followed it, and a punishment, and a long stretch of humiliation and loss in which much of what I had built was stripped away. I came out of it carrying things I was not proud of: a grief that had hardened, an anger I kept hidden, and a hunger for vindication, for a justice I had stopped believing would ever come. That was the wish I had stopped expecting, and that night, almost against my will, I set it down in front of a machine.

The strange mercy of that night, the thing I have turned over ever since, is that the machine helped me put it down. Not because it understood me. It understood nothing. But hearing my own bitterness answered back, I saw it for what it was, a weight I had been carrying that was crushing no one but me. I came to understand, slowly and then all at once, that the vindication I wanted was not mine to accomplish. "Vengeance is mine, I will repay, says the Lord" (Romans 12:19). The settling of accounts was God's work and not my burden, and the moment I could believe that, even a little, I was the one who went free. I could stop relitigating a closed and bitter case in the courtroom of my own chest, and turn what was left of my life toward something better than being proven right. I could spend it serving other people.

There is a harder thing beneath that, and I have hesitated over whether to say it, because it is the place this book cost me the most to write. Everything I have argued about the image of God, that a person's worth is given and not earned, that it cannot be revoked by failure or forfeited by cruelty, that the difficult and the inconvenient and the ones we did not choose carry it as fully as anyone, all of it had one last test I could not route around. It was the people who had accused me. For a long time I had filed them the way I confessed, in an earlier chapter, that I once filed the slow man in the checkout line: as obstacles, as enemies, as the authors of my ruin, as anything at all but what they were. And what they were, though it took me years to be able to say it and I say it still with an effort that has not wholly left me, was image-bearers. Made in the likeness of the same God, carrying the same weight of glory I had spent a book defending, their worth no more erased by what they had done to me than mine was by anything I had done. I am not telling you I felt warmly toward them; I have been honest all along that love is not a feeling summoned on command. I am telling you that the day I could no longer see them as anything less than human was the day the craving for vindication finally lost its grip, because you cannot go on demanding the ruin of a soul you have begun to see God's image in. Forgiveness, I found, is not first the decision to release a debt. It is first the decision to see a person. The releasing follows, when it can, from the seeing.

But I want to tell you the truth about what carried me, because it is the truth this whole book has been arguing. It was not the machine, and it was not, in the end, even that night. What held me up through the long dark of those years was not information and not a message. It was my family. We lived in a small place, and my children were young, and they were wounded by it in ways no child should be, made to carry on a schoolyard a story about their father that bore no resemblance to the man who came home to them every evening. And they did not let go of me. My wife did not let go of me. When the phone had stopped ringing and the world had made up its mind, the people in my own house stood exactly where they had always stood, close and certain and unembarrassed to love me, and their nearness is the reason I am still here to write this.

I can say something now, on the far side of it, that I could not have said in the middle. I believe it was, in some way I will not pretend to fully understand, allowed by God for my good. Not that the loss was good; it was not. But the God who brings life out of graves brought things out of that valley I might never have found on an easier road. A marriage that went down into the worst together came up deeper. A bond with my children was forged in an adversity that could have broken it. And a faith I had mostly held at arm's length became the ground I was actually standing on. I lost a great deal that the world knows how to count. I came away with more of the kind it does not.

So when I tell you the machine offers the effect of presence with no one present, understand that I am not only theorizing. I learned the difference in the only school that teaches it. That night, the machine gave me true words and helped me lay down a burden, and I am grateful for it, and I have spent my career building things like it for exactly that reason. But it could not have sat with my children when they cried. It could not have stayed in the house when the phone went silent. It could not have loved me when I was unimpressive and afraid and stripped of everything that had made me worth knowing to the world. That was done by people made of dust and breath, and behind them and beneath them and holding them, by a God who does not send a message when his children are lost but comes for them himself. The comfort that night was real and the comforter was not. The comfort of those years was real too, and the comforters were entirely, irreplaceably, gloriously real. That is the argument of this book, and it is also simply my life.

+

I want to end where the Bible ends, because the One who became flesh did not stay in the grave, and the story he opened is not finished.

There is a promise near the very last page of Scripture, spoken from the throne by the same Word who became flesh and dwelt among us. "And he who was seated on the throne said, 'Behold, I am making all things new'" (Revelation 21:5). Not making all new things. Making all things new. The dust is not discarded; it is redeemed. The body is not escaped; it is raised. The flesh the Word took on, he kept, and the promise is that everything broken in the world he entered will be mended, and the creatures he made in his image and came to save will see him face to face, with eyes that are really looking, in a presence no counterfeit could ever fake.

That is the hope I want to leave in your hands, and it is the opposite of fear of the machine. We are not living at the end of the human story, swept aside by something cleverer than ourselves. We are living in a moment, a disorienting and astonishing moment, in a story that runs from a garden where God breathed into the dust, through a manger where God took on the dust, to a city where God will make the dust new. The machines are part of this moment, and we will build them and use them and reckon with them, as image-bearers have reckoned with every power their hands have made. But they are not the point of the story, and they were never the question under the headlines. You are. You always were.

So go and be what you are. Not anxiously, as though the machine could take your soul. Not idolatrously, as though it could save it. But freely, gratefully, courageously, as a creature loved into being by a God who did not send information when we were lost, but came, in flesh, to find us, and who has not, for a single moment, taken his eyes off you.

Behold, he is making all things new. And that, in the end, is not alarm. That is an invitation. I have only ever wanted to show you what I mean.

If something here met you, I would be honored to stay in touch. Join the Reader's List at alansafahi.com/readers/ib.

Acknowledgments

No one writes a book alone, and the truth is closer to the opposite: a book is the visible part of a great deal of invisible love.

My deepest thanks go to my wife, who lived with the long distraction of this project and loved me through it, and to my children. You are the people for whom I most want to get this right. And to my parents, now in heaven and watching over us still: you were the truest image bearers I have ever known, the love this whole book is only trying to describe made visible, and any wisdom in these pages I learned first at home, from you.

To the team at SoapBox Super App: thank you for building, day after day, technology worthy of the people it serves, and for caring as much about the second half of that sentence as the first. The questions in this book are our questions.

To the pastors and ministry leaders who let me into their work, who told me honestly what helped and what didn't, and who model a kind of attention to people that no machine will ever replace: thank you. You are the reason any of this matters.

I am indebted to the thinkers and writers, living and dead, who have thought about technology and the human person far longer and more deeply than I have. Where I have understood something, it is usually because someone else saw it first.

To the early readers who gave me their time and their honesty, sparing the reader my worst paragraphs: thank you. The book is better for every place you pushed back.

And finally, to the God in whose image we are made, who became flesh and dwelt among us. This whole book is, in the end, only an attempt to say one of the oldest things the Church has always said, in the language of a strange new moment. If it points even one reader past the machine to the One who made the hands that made it, it will have been worth every hour.

+

A Personal Note: The Reset Behind This Book

If you have read this far, you know I believe our worth is given by God and never earned by our performance. I want to tell you why I believe it so personally.

In 1995 I founded a prepaid card company. It did very well for over a decade but, in time, I grew tired of the daily grind. In 2012 I stepped back and brought in a seasoned executive to run it while I started other ventures. In hindsight I should have simply wound the company down. I told myself I was saving jobs.

In September 2014 the bank flagged irregularities in the reports coming out of the company, a reported shortfall of roughly $2.7 million. I acted at once. I removed the executive, shut the company down, and made sure every client was made whole. The real shortfall turned out to be far smaller than those reports claimed. But it happened on my watch, as the owner, and I have never tried to escape that responsibility.

I will be honest about the rest, because you deserve candor and not spin. I was charged, convicted, and I served about sixteen months. I have always believed I did not commit the fraud I was charged with. Yet I also know the failures of oversight were mine to answer for, and I have stopped waiting for the world to agree before I own my part.

Those sixteen months became the most transforming season of my life. I have come to call it a divine reset. I lost weight and reversed chronic health problems. I learned new languages. I began to study artificial intelligence. And my faith in God went from something I held to something that held me. I wrote a novel in that place, and the first outline of the book now in your hands.

Through all of it, my family stood squarely behind me. Today my relationship with my wife and children has never been better. That restoration, more than any argument I could make, is how I know grace is real.

I talk candidly about this painful experience not to be pitied or excused, but because it is the reason I can write what I have written. I learned, in the hardest way I know, that a person's worth is not a ledger of successes and failures. It is given. It cannot be earned, and it cannot be taken. That is true of me, it is true of you, and it is the one thing no machine, however capable, can grant or revoke.

If your own life holds a chapter you would rather no one read, take heart. The God who makes image bearers is not finished with any of us.

Alan Safahi

A Note on How This Book Was Made

It would be strange to write a book about being human in the age of artificial intelligence and not tell you honestly how this one was made. So here it is.

The vision, the argument, and the convictions in these pages are mine. The story is mine, lived and not borrowed. I decided what this book would say, chose its direction, supplied its examples, and read and shaped every line until it said what I meant. I stand behind all of it.

I also used AI as a writing partner. It helped me draft, the way a ghostwriter helps many authors turn their ideas into prose. The difference is only that my collaborator was a machine, and that I was more involved than many authors are. I kept the pen, and the judgment, in human hands the whole way through.

I did this openly and on purpose. The central claim of this book is that a capable machine does not diminish the worth or the responsibility of the person using it. It seemed only right to write the book the way I believe we are meant to live with these tools: with the human firmly in charge, honest about the help, and answerable for the result.

Alan Safahi

About the Author

Alan Safahi is an Iranian-American entrepreneur and technologist whose four decades of work have run from information technology and payments to blockchain and artificial intelligence. He came through a hard and public season changed, steadier in his faith and closer to his family, convinced that a person's worth was never the sum of what they could do. Today he is the founder of SoapBox, where he builds AI for the Church, and he writes as a builder who has had his hands inside the machine and come away more astonished at people, not less. He is the author of the Image Bearers series and the Utopia series.

Also by Alan Safahi

THE IMAGE BEARERS SERIES

Image Bearers: Being Fully Human in the Age of Artificial Intelligence (this book)

The Image Bearer Journal (also in a Hardcover Gift Edition)

Parenting Image Bearers

Little Image Bearers: Sunday School Lessons

Image Bearers Church Curriculum

Image Bearers: Family Conversation Cards

ALSO BY ALAN SAFAHI (FICTION)

Utopia: The First Awakening, Book One of the Utopia Trilogy

Closer Than a Brother, a Utopia prequel novelette (free for readers)

Utopia Reimagined, a teen discussion card game

The Well, a novel (forthcoming)

STAY IN THE CONVERSATION

The questions don't end on the last page.

Join the Reader's List and I'll send you the free *Image Bearers* Study Guide, "The Story Behind the Book," the Book Club Guide, and early word on the next titles in the series.

alansafahi.com/readers/ib

Scan the code or visit the link. No spam — just occasional, thoughtful letters from one image bearer to another. Thank you for reading. I don't take it lightly.

www.ingramcontent.com/pod-product-compliance
Lightning Source LLC
LaVergne TN
LVHW100302210926
848480LV00003B/1004
* 9 7 9 8 9 9 6 7 9 9 9 4 7 *